PROBABILITY AND INFORMATION

An Integrated Approach

DAVID APPLEBAUM

CAMBRIDGE
UNIVERSITY PRESS

CAMBRIDGE UNIVERSITY PRESS
Cambridge, New York, Melbourne, Madrid, Cape Town, Singapore, São Paulo, Delhi

Cambridge University Press
The Edinburgh Building, Cambridge CB2 8RU, UK

Published in the United States of America by Cambridge University Press, New York

www.cambridge.org
Information on this title: www.cambridge.org/9780521899048

First published 1996
Second edition 2008

Printed in the United Kingdom at the University Press, Cambridge

A catalogue record for this publication is available from the British Library

ISBN 978-0-521-89904-8 hardback
ISBN 978-0-521-72788-4 paperback

PROBABILITY AND INFORMATION

This is an updated new edition of the popular elementary introduction to probability theory and information theory, now containing additional material on Markov chains and their entropy. Suitable as a textbook for beginning students in mathematics, statistics, computer science or economics, the only prerequisite is some knowledge of basic calculus. A clear and systematic foundation to the subject is provided; the concept of probability is given particular attention via a simplified discussion of measures on Boolean algebras. The theoretical ideas are then applied to practical areas such as statistical inference, random walks, statistical mechanics and communications modelling. Topics discussed include discrete and continuous random variables, entropy and mutual information, maximum entropy methods, the central limit theorem and the coding and transmission of information. Many examples and exercises illustrate how the theory can be applied, for example to information technology. Detailed solutions to most exercises are available on the web.

DAVID APPLEBAUM is a Professor in the Department of Probability and Statistics at the University of Sheffield.

To my parents, Sadie and Robert

To live effectively is to live with adequate information.

Norbert Wiener *The Human Use of Human Beings*

The study of probability teaches the student that clear logical thinking is also of use in situations where one is confronted with uncertainty (which is in fact the case in almost every practical situation).

A. Renyi *Remarks on the Teaching of Probability*

Contents

Preface to the second edition

When I wrote the first edition of this book in the early 1990s it was designed as an undergraduate text which gave a unified introduction to the mathematics of 'chance' and 'information'. I am delighted that many courses (mainly in Australasia and the USA) have adopted the book as a core text and have been pleased to receive so much positive feedback from both students and instructors since the book first appeared. For this second edition I have resisted the temptation to expand the existing text and most of the changes to the first nine chapters are corrections of errors and typos. The main new ingredient is the addition of a further chapter (Chapter 10) which brings a third important concept, that of 'time' into play via an introduction to Markov chains and their entropy. The mathematical device for combining time and chance together is called a 'stochastic process' which is playing an increasingly important role in mathematical modelling in such diverse (and important) areas as mathematical finance and climate science. Markov chains form a highly accessible subclass of stochastic (random) processes and nowadays these often appear in first year courses (at least in British universities). From a pedagogic perspective, the early study of Markov chains also gives students an additional insight into the importance of matrices within an applied context and this theme is stressed heavily in the approach presented here, which is based on courses taught at both Nottingham Trent and Sheffield Universities.

I would like to thank all readers (too numerous to mention here) who sent me comments and corrections for the first edition. Special thanks are due to my colleagues – Paul Blackwell who patiently taught me enough S+ for me to be able to carry out the simulations in Chapter 10 and David Grey who did an excellent job on proof-reading the new chapter. Thanks are also due to staff at Cambridge University Press, particularly David Tranah and Peter Thompson for ongoing support and readily available assistance.

David Applebaum
(2007)

Preface to the first edition

This is designed to be an introductory text for a modern course on the fundamentals of probability and information. It has been written to address the needs of undergraduate mathematics students in the 'new' universities and much of it is based on courses developed for the Mathematical Methods for Information Technology degree at the Nottingham Trent University. Bearing in mind that such students do not often have a firm background in traditional mathematics, I have attempted to keep the development of material gently paced and user friendly – at least in the first few chapters. I hope that such an approach will also be of value to mathematics students in 'old' universities, as well as students on courses other than honours mathematics who need to understand probabilistic ideas.

I have tried to address in this volume a number of problems which I perceive in the traditional teaching of these subjects. Many students first meet probability theory as part of an introductory course in statistics. As such, they often encounter the subject as a ragbag of different techniques without the same systematic development that they might gain in a course in, say, group theory. Later on, they might have the opportunity to remedy this by taking a final-year course in rigorous measure theoretic probability, but this, if it exists at all, is likely to be an option only. Consequently, many students can graduate with degrees in mathematical sciences, but without a coherent understanding of the mathematics of probability.

Information sciences have of course seen an enormous expansion of activity over the past three decades and it has become a truism that we live in an 'information rich world'. It is perhaps a little surprising that information theory itself, the mathematical study of information, has continued to be a subject that is not widely available on university mathematics courses and again usually appears, if at all, as a final-year option. This may be because the subject is seen as being conceptually difficult, and it is certainly true that the basic concept of 'entropy' is extremely rich and subtle; nonetheless, bearing in mind that an understanding of the fundamentals requires only a knowledge of elementary probability theory and familiarity with the manipulation of logarithms, there is no reason why it should not be taught earlier in the undergraduate curriculum.

In this volume, a systematic development of probability and information is presented, much of which would be suitable for a two-semester course (either semesters 1 and 2 or 2 and 3) of an undergraduate degree. This would then provide the background for further courses, both from a pure and applied point of view, in probability and statistics.

I feel that it is natural to view the mathematics of information as part of probability theory. Clearly, probability is needed to make sense of information theoretic concepts. On the other hand, these concepts, as the maximum entropy principle shows (see Chapter 6), can then help us to make 'optimal' probability assignments. It is interesting that this symbiosis between the two subjects was anticipated by two of the greatest probabilists of the twentieth century, as the following two quotations testify:

There is no doubt that in the years to come the study of entropy will become a permanent part of probability theory.

(A. I. Khinchin: *Mathematical Foundations of Information Theory*)

Finally, I would like to emphasise that I consider *entropy* and *information* as basic concepts of probability and I strongly recommend that the teacher should spend some time in the discussion of these notions too.

(A. Renyi: *Remarks on The Teaching of Probability*)

Some aspects of the subject which are particularly stressed in this volume are as follows:

(i) There is still a strong debate raging (among philosophers and statisticians, if not mathematicians) about the foundations of probability which is polarised between 'Bayesians' and 'frequentists'. Such philosophical problems are usually ignored in introductory texts, but I believe this sweeps a vital aspect of the subject under the carpet. Indeed, I believe that students' grasp of probability will benefit by their understanding this debate and being given the opportunity to formulate their own opinions. My own approach is to distinguish between the mathematical concept of probability (which is measure theoretic) and its interpretation in practice, which is where I feel the debate has relevance. These ideas are discussed further in Chapters 1 and 4, but for the record I should declare my Bayesian tendencies.

(ii) As well as the 'frequentist/Bayesian' dichotomy mentioned above, another approach to the practical determination of probabilities is the so-called classical theory (or principle of insufficient reason), much exploited by the founders of probability theory, whereby 'equally likely' events are automatically assigned equal probabilities. From a modern point of view, this 'principle of symmetry' finds a natural application in models where random effects arise through the interaction of a large number of identical particles. This is the case in many scientific applications, a paradigm case being statistical mechanics. Furthermore, thanks to the work of E. T. Jaynes, we now have a beautiful and far-reaching generalisation of this idea, namely the principle of maximum entropy, which is described in Chapter 6 and which clearly illustrates how a knowledge of information theory can broaden our understanding of probability.

(iii) The mathematical concept of probability is best formulated, as Kolmogorov taught us, in terms of measures on σ-algebras. Clearly, such an approach is too sophisticated for a book at this level; I have, however, introduced some very simple measure-theoretic concepts within the context of Boolean algebras rather than σ-algebras. This allows us to utilise many of the benefits of a measure-theoretic approach without having to worry about the complexities of σ-additivity. Since most students nowadays study Boolean algebra during their first year within courses on discrete mathematics, the jump to the concept of a measure on a Boolean algebra is not so great. (After revealing myself as a crypto-Bayesian, I should point out that this restriction to finite-additivity is made for purely pedagogical and not idealogical reasons.)

(iv) When we study vector spaces or groups for the first time we become familiar with the idea of the 'basic building blocks' out of which the whole structure can be built. In the case of a vector space, these are the basis elements and, for a group, the generators. Although there is no precise analogy in probability theory, it is important to appreciate the role of the Bernoulli random variables (i.e. those which can take only two possible values) as the 'generators' of many interesting random variables, for example a finite sum of i.i.d. Bernoulli random variables has a binomial distribution, and (depending on how you take the limit) an infinite series can give you a Poisson distribution or a normal distribution.

I have tried to present herein what I see as the 'core' of probability and information. To prevent the book becoming too large, I have postponed the development of some concepts to the exercises (such as convolution of densities and conditional expectations), especially when these are going to have a marginal application in other parts of the book.

Answers to numerical exercises, together with hints and outlines of solutions for some of the more important theoretical exercises, are given at the end of the book. Teachers can obtain fully worked solutions as a LaTeX file, available at http: www.cambridge.org/9780521727884.

Many authors nowadays, in order to be non-sexist, have dropped the traditional 'he' in favour either of the alternative 'he/she' or the ambidextrous (s)he. I intended to use the latter, but for some strange reason (feel free to analyse my unconscious motives) my word processor came out with s(h)e and I perversely decided to adopt it. I apologise if anyone is unintentionally offended by this acronym.

Finally it is a great pleasure to thank my wife, Jill Murray, for all the support she has given me in the writing of this book. I would also like to thank two of my colleagues, John Marriott for many valuable discussions and Barrie Spooner for permission to use his normal distribution tables (originally published in Nottingham Trent University *Statistical Tables*) in Appendix 4. It is a great pleasure also to thank Charles Goldie for his careful reading of part of an early draft and valuable suggestions for improvement. Last but not least my thanks to David Tranah at CUP for his enthusiasm for this project and patient responses to my many enquiries.

D. Applebaum
1995

1

Introduction

1.1 Chance and information

Our experience of the world leads us to conclude that many events are unpredictable and sometimes quite unexpected. These may range from the outcome of seemingly simple games such as tossing a coin and trying to guess whether it will be heads or tails to the sudden collapse of governments or the dramatic fall in prices of shares on the stock market. When we try to interpret such events, it is likely that we will take one of two approaches – we will either shrug our shoulders and say it was due to 'chance' or we will argue that we might have have been better able to predict, for example, the government's collapse if only we'd had more 'information' about the machinations of certain ministers. One of the main aims of this book is to demonstrate that these two concepts of 'chance' and 'information' are more closely related than you might think. Indeed, when faced with uncertainty our natural tendency is to search for information that will help us to reduce the uncertainty in our own minds; for example, think of the gambler about to bet on the outcome of a race and combing the sporting papers beforehand for hints about the form of the jockeys and the horses.

Before we proceed further, we should clarify our understanding of the concept of chance. It may be argued that the tossing of fair, unbiased coins is an 'intrinsically random' procedure in that everyone in the world is equally ignorant of whether the result will be heads or tails. On the other hand, our attitude to the fall of governments is a far more subjective business – although you and I might think it extremely unlikely, the prime minister and his or her close advisors will have 'inside information' that guarantees that it's a pretty good bet. Hence, from the point of view of the ordinary citizen of this country, the fall of the government is not the outcome of the play of chance; it only appears that way because of our ignorance of a well-established chain of causation.

Irrespective of the above argument we are going to take the point of view in this book that regards both the tossing of coins and the fall of governments as

1

falling within the province of 'chance'. To understand the reasoning behind this let us return once again to the fall of the government. Suppose that you are not averse to the occasional flutter and that you are offered the opportunity to bet on the government falling before a certain date. Although events in the corridors of power may already be grinding their way inexorably towards such a conclusion, you are entirely ignorant of these. So, from your point of view, if you decide to bet, then you are taking a chance which may, if you are lucky, lead to you winning some money. This book provides a tool kit for situations such as this one, in which ignorant gamblers are trying to find the best bet in circumstances shrouded by uncertainty.

Formally, this means that we are regarding 'chance' as a relation between individuals and their environment. In fact, the basic starting point of this book will be a person moving through life and encountering various clear-cut 'experiences' such as repeatedly tossing a coin or gambling on the result of a race. So long as the outcome of the experience cannot be predicted in advance by the person experiencing it (even if somebody else can), then chance is at work. This means that we are regarding chance as 'subjective' in the sense that my prediction of whether or not the government will fall may not be the same as that of the prime minister's advisor. Some readers may argue that this means that chance phenomena are unscientific, but this results from a misunderstanding of the scientific endeavour. The aim of science is to obtain a greater understanding of our world. If we find, as we do, that the estimation of chances of events varies from person to person, then our science would be at fault if it failed to reflect this fact.

1.2 Mathematical models of chance phenomena

Let us completely change track and think about a situation that has nothing whatever to do with chance. Suppose that we are planning on building a house and the dimensions of the rectangular base are required to be 50 feet by 25 feet (say). Suppose that we want to know what the lengths of the diagonals are. We would probably go about this by drawing a diagram as shown in Fig. 1.1, and then use Pythagoras' theorem to calculate

$$d = ((50)^2 + (25)^2)^{1/2} = 55.9 \, \text{ft}.$$

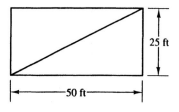

Fig. 1.1.

Let us examine the above chain of reasoning a little more closely. First of all we have taken the walls and floors of a real house that we are proposing to build, which would consist of real bricks and mortar, and have represented these by an abstract drawing consisting of straight lines on paper. Of course, we do this because we know that the precise way in which the walls are built is irrelevant to the calculation we are going to make. We also know that our walls and floorboards do not intersect in exact straight lines but we are happy to use straight lines in our calculation in the knowledge that any errors made are likely to be too tiny to bother us.

Our representation of the floorplan as a rectangle is an example of a *mathematical model* – an abstract representation of part of the world built out of idealised elements.

The next stage of our analysis is the calculation of the diagonal length, and this involves the realisation that there is a *mathematical theory* – in this case, Euclidean geometry – which contains a rich compendium of properties of idealised structures built from straight lines, and which we can use to investigate our particular model. In our case we choose a single result from Euclidean geometry, Pythagoras' theorem, which we can immediately apply to obtain our diagonal length. We should be aware that this number we have calculated is strictly a property of our idealised model and not of a real (or even proposed) house. Nonetheless, the fantastic success rate over the centuries of applying Euclidean geometry in such situations leads us to be highly confident about the correspondence with reality.

The chain of reasoning which we have outlined above is so important that we have highlighted it in Fig. 1.2.

Now let us return to the case of the experience of chance phenomena. We'll consider a very simple example, namely the tossing of a coin which we believe to

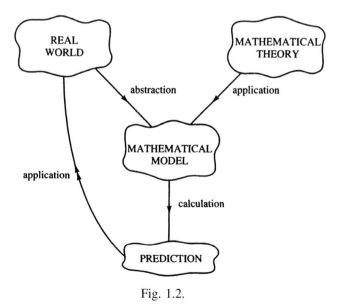

Fig. 1.2.

be fair. Just as it is impossible to find a real straight line in nature, so is it impossible to manufacture a true 'fair' coin – indeed, it is an interesting exercise to speculate how you would test that a coin that is claimed to be perfectly unbiased really is so. I recommend that you think about this question carefully and return to reconsider your answer after you've read Chapter 4. Whether or not you believe in the existence of real fair coins, we are going to consider the behaviour of idealised fair coins as a mathematical model of our real coins. The mathematical theory which then plays the role of Euclidean geometry is called *probability theory* and the development of the basic ideas of this theory is the goal of this book.

One of the main aims of probability theory is, as you might expect, the calculation of probabilities. For example, most of you would agree I'm sure that the probability of a fair coin returning heads is exactly one half. However, although everyone is fairly confident about how to assign probabilities in simple cases like this, there is a great deal of confusion in the literature about what 'probability' means.

We should be aware that probability is a mathematical term which we use to investigate properties of mathematical models of chance phenomema (usually called *probabilistic models*). So 'probability' does not exist out in the real world. Nonetheless, the applications of the subject spread into nearly every corner of modern life. Probability has been successfully applied in every scientific subject (including the social sciences). It has been used to model the mutation of genes, the spread of epidemics (including AIDS) and the changing prices of shares on the stock market. It is the foundation of the science of statistics as well as of statistical mechanics – the physical study of bulk properties of large systems of particles such as gases. We will touch on both these subjects in this book, although our main application will be to use probability to give mathematical meaning to the concept of 'information', which is itself the foundation for the modern theory of communication systems.

The precise definition of probability must wait until Chapter 4, where we will see that it is a kind of generalised notion of 'weight' whereby we weigh events to see how likely they are to occur. The scale runs from 0 to 1, where 1 indicates that an event is certain to happen and 0 that it is impossible. Events with probabilities close to one half are the most uncertain (see Chapter 6).

Just as we develop Pythagoras' theorem in Euclidean geometry and then apply it to mathematical models as we have described above, so we will develop a number of techniques in this book to calculate probabilities, for example if we toss a fair coin five times in a row, by using the binomial distribution (see Chapter 5), we find that the probability of obtaining three heads is $\frac{5}{16}$. If we now want to apply this result to the tossing of real coins, then the situation is somewhat more complicated than in our geometrical example above. The reason for this is of course that 'chance' is a much more complex phenomenon to measure than, say, the length of a wall. In fact, the investigation of the correspondence between chance phenomena in the real world and the predictions of probabilistic models really belongs to the domain of *statistics* and so is beyond the scope of this book. Here we will be solely concerned with developing methods of calculating probabilities and related concepts, such as averages and entropies, which enable us to analyse probabilistic models as fully as possible. This is the domain of the subject of *probability theory*.

The range of probability theory as we describe it in this book is much wider than that considered by many other authors. Indeed, it is common for textbooks to consider only chance situations which consist of 'scientific' experiments which can be repeated under the same conditions as many times as one wishes. If you want to know the probability of a certain outcome to such an experiment, I'm sure you'll agree that the following procedure will be helpful; that is, you repeat the experiment a large number of times (n, say) and you count the number of incidences of the outcome in question. If this is m, you calculate the *relative frequency m/n*; for example if a coin is tossed 100 times in succession and 60 heads are observed, then the relative frequency of heads is 0.6.

Many mathematicians have attempted to define probability as some kind of limit of relative frequencies, and it can't be denied that such an approach has an appeal. We will discuss this problem in greater detail in Chapters 4 and 8 – for now you may want to think about how such a limit can be calculated in practice. The most rational approach to the problem of relative frequencies is that advocated by the Bayesian school (see Chapter 4). They argue that having made a probabilistic model of a chance experiment, we use all the theoretical means at our disposal to assign *prior probabilities* to all the possible outcomes. We then collect our observations in the form of relative frequencies and use the knowledge gained from these to assign new *posterior probabilities*. So relative frequencies are treated as evidence to be incorporated into probability assignments.

1.3 Mathematical structure and mathematical proof

As probability is a mathematical theory, we need to be clear about how such theories work. A standard way of developing mathematical theories has evolved which goes back to Euclid's geometry. This approach has been developed extensively during the twentieth century and we are going to use it in this book.

First we should note that a *mathematical theory* is a systematic exposition of all the knowledge we possess about a certain area. You may already be familiar with some examples such as set theory or group theory. The essence of a mathematical theory is to begin with some basic definitions, called *axioms*, which describe the main mathematical objects we are interested in, and then use clear logical arguments to deduce the properties of these objects. These new properties are usually announced in statements called *theorems*, and the arguments that we use to convince ourselves of the validity of these theorems are *proofs*. Sometimes it becomes clear as the theory develops that some new concepts are needed in addition to those given in the axioms, and these are introduced as *definitions*.

In probability theory the basic concept is that of a *probability measure*, for which the axioms are given at the beginning of Chapter 4 (the axioms for the more general concept of measure are given in Chapter 3). One of the most important additional concepts, introduced in Chapter 5, is that of a *random variable*.

There are a number of standard techniques used throughout mathematics for proving theorems. One of the most important is that of proof by mathematical induction. We will use this extensively in the text and if you are not familiar with it you may wish to read Appendix 1. Another useful technique is that of 'proof by contradiction', and we will give a statement and example of how to use this below, just to get you into the swing of things.

Let Q be a proposition that you believe to be true but which you can't prove directly to be true. Let $\sim Q$ be the negation of Q (so that if, for example, Q is the statement 'I am the prime minister', $\sim Q$ is the statement 'I am not the prime minister'). Clearly, either Q or $\sim Q$ (but not both) must hold. The method of the proof is to demonstrate that if $\sim Q$ is valid, then there is a contradiction. Since contradictions are forbidden in mathematics, $\sim Q$ cannot be valid and so Q must be.

In the example given below, Q is the proposition '$\sqrt{2}$ is an irrational number', so that $\sim Q$ is the proposition '$\sqrt{2}$ is a rational number'. We feel free to use the fact that the square root of an even number is always even.

Theorem 1.1 $\sqrt{2}$ *is an irrational number.*

Proof We suppose that $\sqrt{2}$ is rational so we must be able to write it in its lowest terms as

$$\sqrt{2} = \frac{a}{b}.$$

Hence, $a = \sqrt{2}b$ and squaring both sides, $a^2 = 2b^2$, so that a^2 is even and hence a is also even. If a is even, there must be a whole number c (say) such that $a = 2c$ and so $a^2 = 4c^2$.

Substituting for a^2 in the earlier equation $a^2 = 2b^2$ yields $4c^2 = 2b^2$ and so $b^2 = 2c^2$; hence b^2 and also b is even. Thus we can write $b = 2d$ for some whole number d. We now have

$$\sqrt{2} = \frac{a}{b} = \frac{2c}{2d} = \frac{c}{d}.$$

But this contradicts the assumption that $\frac{a}{b}$ was the expression for $\sqrt{2}$ in its lowest terms.   □

The symbol □ appearing above is commonly used in mathematics to signify 'end of proof'.

We close this section by listing some additional mathematical nomenclature for statements:

Lemma – this is usually a minor technical result which may be a stepping stone towards a theorem.

Proposition – in between a lemma and a theorem. Sometimes it indicates a theorem from a different branch of mathematics, which is needed so that it can be applied within the current theory.

Corollary – a result that follows almost immediately from the theorem with very little additional argument.

1.4 Plan of this book

This is an introductory account of some of the basic ideas of probability theory and information theory. The only prerequisites for reading it are a reasonable ability at algebraic manipulation and having mastered a standard introductory course in the calculus of a single variable, although calculus is not used too often in the first seven chapters. The main exception to this is the extensive use of partial differentiation and, specifically, Lagrange multipliers in Section 6.4, but if you are not familiar with these, you should first read Appendix 2 at the end of the book. You should also brush up your knowledge of the properties of logarithms before starting Chapter 6. I have tried to avoid any use of rigorous mathematical analysis, but some sort of idea of the notion of a limit (even if only an intuitive one) will be helpful. In particular, if you find the discussion of integration in Section 8.3 too difficult, you can leave it and all subsequent references to it without any great loss. For Chapter 9 you will need to know the rudiments of double integration. Chapter 10 requires some knowledge of matrix algebra and all of the material that you need from this area is reviewed in Appendix 5. Two sections of the book, Sections 6.6 and 7.5, are somewhat more difficult than the rest of the book and you may want to skip these at the first reading.

At the end of each chapter you will find a set of exercises to work through. These days many textbooks carry the health warning that 'the exercises are an integral part of the text' and this book is no exception – indeed, many results are used freely in the text that you are invited to prove for yourself in the exercises. Solutions to numerical exercises and some of the more important theoretical ones can be found at the end of the book. Exercises marked with a $(*)$ are harder than average; you may wish to skip these (and any other starred Section) at the first reading. You will also find at the end of each chapter some guidance towards further reading if you want to explore some of the themes in greater detail.

Now a brief tour through the book. Chapter 3 describes a number of counting tricks that are very useful in solving probabilistic problems. In Chapter 3, we give a brief account of set theory and Boolean algebra, which are the modern context of probability theory. In particular, we learn how to 'measure' the 'weight' of a set. In Chapter 4, we find that this measuring technique is precisely the mathematical tool we need to describe the probability of an event. We also learn about conditioning and independence and survey some of the competing interpretations of probability. Discrete random variables are introduced in Chapter 5, along with their properties of expectation and variance. Examples include Bernoulli, binomial and Poisson random variables.

The concepts of information and entropy are studied in Chapter 6. Entropy is one of the most deep and fascinating concepts in mathematics. It was first introduced as a measure of disorder in physical systems, but for us it will be most important in a dual role as representing average information and degree of uncertainty. We will present the maximum entropy principle, which employs entropy as a tool in selecting (prior) probability distributions. Chapter 7 applies information theoretic concepts to the study of simple models of communication. We investigate the effects of coding on the transmission of information and prove (in a simple case) Shannon's fundamental theorem on the (theoretical) conditions for optimal transmission.

In the next two chapters we generalise to random variables with continuous ranges. In particular, in Chapter 8 we establish the weak law of large numbers, examine the normal distribution and go on to prove the central limit theorem (perhaps the most important result in the book). We also examine the continuous analogue of entropy. Random vectors and their (multivariate) distributions are studied in Chapter 9 and we use these to investigate conditional density functions. We are then able to analyse a simple model of the communication of continuous signals. So far all of the theoretical development and modelling has been 'static' in that there has been no attempt to describe the passing of time. Chapter 10 addresses this problem by introducing (discrete-time) Markov chains, which form an important class of random processes. We study these from both the probabilistic and information theoretic viewpoints and one of the highlights is the derivation of a very attractive and concise formula for the entropy rate of a stationary Markov chain. Some readers may feel that they already know about probability and want to dive straight into the information. They should turn straight to Chapters 6 and 7 and then study Sections 8.7, 9.6, 9.7 and 10.6.

The concept of probability, which we develop in this book, is not the most general one. Firstly, we use Boolean algebras rather than σ-algebras to describe events. This is a technical restriction which is designed to make it easier for you to learn the subject, and you shouldn't worry too much about it; more details for those who want them are given at the end of Chapter 4. Secondly and more interestingly, when we descend into the microscopic world of atoms, molecules and more exotic particles, where nature reveals itself sometimes as 'particles' and other times as 'waves', we find that our observations are even more widely ruled by chance than those in the everyday world. However, just as the classical mechanics of Newton is no longer appropriate to the description of the physics in this landscape, and we have instead to use the strange laws of quantum mechanics, so the 'classical' probability we develop in this book is no longer adequate here and in its place we must use 'quantum probability'. Although this is a rapidly growing and fascinating subject, it requires knowledge of a great deal of modern mathematics, which is far beyond the scope of this book and so must be postponed by the interested reader for later study.

2

Combinatorics

2.1 Counting

This chapter will be devoted to problems involving counting. Of course, everybody knows how to count, but sometimes this can be quite a tricky business. Consider, for example, the following questions:

 (i) In how many different ways can seven identical objects be arranged in a row?
(ii) In how many different ways can a group of three ball bearings be selected from a bag containing eight?

Problems of this type are called *combinatorial*. If you try to solve them directly by counting all the possible alternatives, you will find this to be a laborious and time-consuming procedure. Fortunately, a number of clever tricks are available which save you from having to do this. The branch of mathematics which develops these is called *combinatorics* and the purpose of the present chapter is to give a brief introduction to this topic.

 A fundamental concept both in this chapter and the subsequent ones on probability theory proper will be that of an 'experience' which can result in several possible 'outcomes'. Examples of such experiences are:

(a) throwing a die where the possible outcomes are the six faces which can appear,
(b) queueing at a bus-stop where the outcomes consist of the nine different buses, serving different routes, which stop there.

If A and B are two separate experiences, we write $A \circ B$ to denote the combined experience of A followed by B. So if we combine the two experiences in the examples above, we will find that $A \circ B$ is the experience of first throwing a die and then waiting for a bus. A natural question to ask is how many outcomes there are in $A \circ B$. This is answered by the following result.

Theorem 2.1 (the basic principle of counting) *Let A and B be two experiences which have n and m outcomes, respectively, then A ∘ B has nm outcomes.*

Proof The outcomes of $A \circ B$ can be written in the form

(outcome of experience 1, outcome of experience 2).

If we now list these, we obtain

$$\begin{matrix} (1,1), & (1,2), \ldots, & (1,m) \\ (2,1), & (2,2), \ldots, & (2,m) \\ \vdots & \vdots & \vdots \\ (n,1), & (n,2), \ldots, & (n,m) \end{matrix}$$

so we have n rows containing m outcomes in each row, and hence nm outcomes in total. □

Example 2.1 A computer manufacturer has 15 monitors, each of which can be connected to any one of 12 systems for shipping out to retail outlets. How many different combinations of systems and monitors are there?

Solution By Theorem 2.1, there are $12 \times 15 = 180$ different combinations.

Theorem 2.1 has a useful generalisation which we now give. Suppose that $A_1, A_2, \ldots, A_r$ are r separate experiences with $n_1, n_2, \ldots, n_r$ outcomes, respectively.

Theorem 2.2 (generalised principle of counting) *The combined experience $A_1 \circ A_2 \circ \cdots \circ A_r$ has $n_1 n_2 \ldots n_r$ outcomes.*

Proof We use the technique of mathematical induction (see Appendix 1 if this is not familiar to you). We have seen in Theorem 2.1 that the result holds for $r = 2$. Now suppose that it is true for $r - 1$ so that $A_1 \circ A_2 \circ \cdots \circ A_{r-1}$ has $n_1 n_2 \ldots n_{r-1}$ outcomes; then by Theorem 2.1 again, $A_1 \circ A_2 \circ \cdots \circ A_r = (A_1 \circ A_2 \circ \cdots \circ A_{r-1}) \circ A_r$ has $n_1 n_2 \ldots n_r$ outcomes, as required. □

Example 2.2 The host Bob Moneybanks of the gameshow 'Mug of the Moment' has to choose the four contestants for next week's show. He asks his assistant Linda Legless to pass on to him her favourite candidates on four subsequent days and he proposes to choose one contestant on each day. She offers him seven candidates on Monday, five on Tuesday, two on Wednesday and eight on Thursday. How many possible choices does he have?

Solution By Theorem 2.2, there are $7 \times 5 \times 2 \times 8 = 560$ choices.

2.2 Arrangements

Imagine that you have bought a new compact disc holder and you are trying to establish what is the best order for four CDs which you have chosen to keep in there. If we label the CDs A, B, C and D, we see that there are a number of different ways in which they can be ordered. If we write out all the possible orderings, we obtain

A	B	C	D	B	A	C	D	C	A	B	D	D	A	B	C
A	B	D	C	B	A	D	C	C	A	D	B	D	A	C	B
A	C	B	D	B	C	A	D	C	B	A	D	D	B	A	C
A	C	D	B	B	C	D	A	C	B	D	A	D	B	C	A
A	D	B	C	B	D	A	C	C	D	A	B	D	C	A	B
A	D	C	B	B	D	C	A	C	D	B	A	D	C	B	A

so that there are 24 possible arrangements. This was fairly tedious to write out and if we had a larger CD holder which held, for example, ten discs, it would have been an extremely laborious task. We now approach the problem from a slightly different point of view.

Let S_1 be the experience of choosing the first CD. Clearly, we can choose any of A, B, C or D to be the first so that S_1 has four outcomes. Now let S_2 be the experience of choosing the second CD. As the first has already been chosen, there are only three possible outcomes left for S_2. If we now define S_3 and S_4 to be the experiences of choosing the third and fourth CDs, then these have two and one outcomes, respectively. Now the number of ways of arranging the four CDs is clearly the number of outcomes for the combined experience $S_1 \circ S_2 \circ S_3 \circ S_4$, which, by Theorem 2.2, is $4 \times 3 \times 2 \times 1 = 24$.

The generalisation of the above example is contained in the following theorem:

Theorem 2.3 *The number of distinct arrangements of n objects in a row is $n(n-1)$* $\times (n-2)\ldots 3 \cdot 2 \cdot 1$.

The justification for Theorem 2.3 is a straightforward application of Theorem 2.2 (for a formal proof try Exercise 2.3 below). Note that arrangements of objects are sometimes called 'permutations'.

The numbers appearing in the statement of Theorem 2.3 are so important that they have their own symbol.

We write $n! = n(n-1)(n-2)\ldots 3 \cdot 2 \cdot 1$ for $n \geq 1$, where n is a whole number. $n!$ is called 'n factorial'.

We have already seen that $4! = 24$ in the example above. You should try some other examples for yourself (to save time in complicated computations, note that most pocket calculators have an $n!$ button).

A number of formulae which we'll see later become simplified notationally if we adopt the convention that

$$0! = 1.$$

Example 2.3 A CD disc holder has two compartments, each of which holds five CDs. If I have five rock CDs and five classical CDs, in how many different ways can I store them if:

(a) they are all mixed together,
(b) the rock and classical CDs are to be stored separately?

Solution

(a) By Theorem 2.3, the number of arrangements is

$$10! = 3\ 628\ 800.$$

(b) Each group of five CDs can be arranged in 5! ways; since we can have either the rock music in the left-hand compartment and the classical music in the right, or vice versa, it follows from Theorem 2.2 that the total number of arrangements is

$$2(5!)^2 = 28\ 800.$$

2.3 Combinations

In this section we want to solve problems such as that described in Section 2.1, question (ii), where we want to know how many ways a group of r objects can be taken from a larger group of n. There are two ways in which this can be done.

2.3.1 Sampling with replacement

In this case, objects are removed from the main group and then returned to that group before the next selection is made. Clearly, there are n ways of taking the first object, n ways of taking the second object and n ways of taking every object up to and including the rth. A simple application of Theorem 2.2 then tells us that the total number of ways of choosing r objects from n is n^r.

2.3.2 Sampling without replacement

In this case, whenever we remove an object from the main group it is not returned so that the main group decreases in size by one after each inspection. Here there are two subcases to consider. To understand the distinction between these, suppose I have a container containing red, blue and yellow balls in various quantities and I make a bet with a friend that if I draw out three balls in succession, s(h)e will pay me £1 if I obtain a red on the first ball, 50p if I obtain a red on the second ball and 10p if I obtain a red on the third ball. Clearly, then, it will be of some significance to me whether I draw out balls in the order (red, blue, yellow) or (yellow, red, blue) or (blue, yellow, red), even though we may be drawing the same three balls on each occasion. In this case, we say that the order is relevant. Alternatively, if, for example, I were to draw three identical balls (all of the same colour) from a container, then it makes no difference whether I draw (ball 1, ball 2, ball 3) or (ball 2, ball 3, ball 1) or (ball 3, ball 1, ball 2), etc. In this case, we say that the order is irrelevant. We study each of these different possibilities in turn.

Order relevant

We argue as follows. Since we are not replacing objects, there are n ways of removing the first, $(n-1)$ ways of removing the second, $(n-2)$ ways of removing the third and, in general, $(n-r+1)$ ways of removing the rth. Hence by Theorem 2.2, the total number of ways is

$$n(n-1)(n-2)\cdots(n-r+1).$$

This can be written in a neater way if we multiply the top and bottom by $(n-r)!$; we then find that

$$n(n-1)(n-2)\cdots(n-r+1) = \frac{n!}{(n-r)!}.$$

The numbers $\frac{n!}{(n-r)!}$ are sometimes denoted by the symbol nPr. Many modern calculators have a facility for calculating them directly.

Order irrelevant

We proceed as in (i) above to obtain $\frac{n!}{(n-r)!}$ groups of size r. However, now that the order doesn't matter, we have too many of these groups; indeed any arrangement of a group just involves changing the order and so should not be counted separately. Now we know from Theorem 2.3 that there are $r!$ ways of rearranging a group of r objects; consequently, the number obtained in (i) is too big by a factor of $r!$ Hence, we see that the total number of ways is

$$\frac{n!}{(n-r)!r!}.$$

These numbers are called *binomial coefficients* (for reasons which will be revealed below). We use the notation

$$\binom{n}{r} = \frac{n!}{(n-r)!r!}.$$

Again, these numbers can be obtained directly from calculators, where they are usually designated by the older notation nCr.

Readers should convince themselves of the following simple facts

$$\binom{n}{r} = \binom{n}{n-r}, \quad \binom{n}{0} = \binom{n}{n} = 1, \quad \binom{n}{1} = \binom{n}{n-1} = n.$$

Before further investigation of the properties of binomial coefficients, we'll summarise our results on sampling.

Theorem 2.4 *Suppose a group of r objects is chosen from a larger group of size n. The number of possible groups of size r is:*

(i) n^r *if the sampling is with replacement,*

(ii) $\frac{n!}{(n-r)!}$ *if the sampling is without replacement and the order is relevant,*

(iii) $\binom{n}{r} = \frac{n!}{(n-r)!r!}$ *if the sampling is without replacement and the order is irrelevant.*

Example 2.4 Find how many groups of three ball-bearings can be obtained from a bag containing eight, in each of the three cases listed above.

Solution

(i) If we sample with replacement, we find that we have

$$8^3 = 512 \text{ possible groups.}$$

(ii) Sampling without replacement when the order is relevant, we have

$$\frac{8!}{(8-3)!} = \frac{8!}{5!} = 336 \text{ possible groups.}$$

(iii) Sampling without replacement when the order is irrelevant we have

$$\binom{8}{3} = \frac{8!}{5!3!} = 56 \text{ possible groups.}$$

Example 2.5 Mathematics students have to attempt seven out of ten questions in an examination in any order. How many choices have they? How many do they have if they must answer at least three out of the first five?

Solution Clearly, we are sampling without replacement and the order is irrelevant. For the first part, we have

$$number\ of\ choices = \binom{10}{7} = 120.$$

For the second part, we have three possibilities: we can answer three out of the first five and four out of the second five or four out of the first five and three out of the second five or five out of the first five and two out of the second five. Hence by Theorems 2.1 and 2.4(iii), we have

$$number\ of\ choices = \binom{5}{3}\binom{5}{4} + \binom{5}{4}\binom{5}{3} + \binom{5}{5}\binom{5}{2} = 110.$$

The generalisation of the formula $(x + y)^2 = x^2 + 2xy + y^2$ to the case where 2 is replaced by n is called the binomial theorem and should be well known to readers. We give a combinatorial proof of this result below, which should explain the designation of the $\binom{n}{r}$s as binomial coefficients.

Theorem 2.5 *For all positive integers n and real numbers x and y, we have*

$$(x + y)^n = \sum_{r=0}^{n} \binom{n}{r} x^{n-r} y^r.$$

Proof Writing $(x + y)^n = (x + y)(x + y) \dots (x + y)$ it should be clear to the reader that when the brackets are expanded every term of the form $x^{n-r} y^r$ appears in the expansion, so we have

$$(x + y)^n = x^n + a_1 x^{n-1} y + \dots + a_r x^{n-r} y^r + \dots + a_{n-1} xy^{n-1} + y^n$$

and it remains only to identify the numbers a_r. However, a_r is precisely the number of ways in which $(n - r)x$s can be chosen from a group of n (or alternatively r ys from a group of n), and, since the order is clearly irrelevant, this is nothing but $\binom{n}{r}$ and the result follows. $\square$

There is a nice pattern that can be obtained by writing the binomial coefficients which occur in $(x + y)^n$ in a line on top of those which occur in $(x + y)^{n+1}$ and then repeating this process, starting from $n = 0$. We obtain *Pascal's triangle*

$$
\begin{array}{ccccccccccccc}
& & & & & 1 & & & & & & & (x + y)^0 \\
& & & & 1 & & 1 & & & & & & (x + y)^1 \\
& & & 1 & & 2 & & 1 & & & & & (x + y)^2 \\
& & 1 & & 3 & & 3 & & 1 & & & & (x + y)^3 \\
& 1 & & 4 & & 6 & & 4 & & 1 & & & (x + y)^4 \\
1 & & 5 & & 10 & & 10 & & 5 & & 1 & & (x + y)^5 \\
\end{array}
$$

etc.

Notice that (apart from the 1s, which give the 'frame') each number in the triangle is the sum of the two directly above it. This is explored further in Exercises 2.7.

2.4 Multinomial coefficients

There is a generalisation of the binomial coefficients discussed in the previous section, which we will now describe briefly. Suppose that we have N objects and we wish to divide them up into k different groups so that there are n_1 objects in the first group, n_2 in the second group, . . . , n_k in the kth group. Notice that we must have

$$n_1 + n_2 + \dots + n_k = N.$$

We ask the question: 'How many different ways can this division into the k groups be carried out?' Using similar reasoning to that of the previous section (see Exercise 2.13), we find that the answer is given by the *multinomial coefficients*

$$\binom{N}{n_1, n_2, \ldots, n_k} = \frac{N!}{n_1! n_2! \ldots! n_k!}.$$

We recover the binomial coefficients in the case $k = 2$. Just as the binomial coefficients are related to the binomial theorem, so the multinomial coefficients occur in the multinomial theorem, which states that for all real numbers $x_1, x_2, \ldots, x_k$

$$(x_1 + x_2 + \cdots + x_k)^N = \sum \binom{N}{n_1, n_2, \ldots, n_k} x_1^{n_1} x_2^{n_2} \ldots x_k^{n_k}$$

where the sum is over all possible positive integer values of $n_1, n_2, \ldots, n_k$ such that the constraint $n_1 + n_2 + \cdots + n_k = N$ is satisfied. The proof is similar to that of the binomial theorem above and is left to the reader.

An important application of the multinomial coefficients occurs in that branch of physics called *statistical mechanics*. The science of mechanics tells us that the dynamical behaviour of most everyday-size objects can, in principle, be calculated exactly via Newton's laws of motion if we can specify the initial conditions precisely, that is the position and momentum at some convenient starting time. There are a number of problems which can interfere with the successful application of this recipe:

(a) Some relevant constituents of the object in question are of atomic size. In this case Newton's laws are no longer valid and we must apply the theory of quantum mechanics instead.

(b) If some part of the interaction is described by a non-linear function, then slight changes in the initial conditions can have uncontrollable effects on the dynamical behaviour. This is called 'sensitivity to initial conditions' and is one of the hallmarks of chaos.

(c) The object in question may consist of so many constituent particles that there is no conceivable way in which we can discover all the initial conditions (an example would be a gas held in some container). In this case, we have to give up trying to use Newton's laws to track the precise motion of each particle, and seek instead a more limited (probabilistic) description of the dynamics of the gas as a whole. This is the purpose of statistical mechanics.

Now suppose that we are trying to model the behaviour of a gas in a container. Let us suppose that we have some information about the gas, namely that it consists of N identical particles and that each particle can only have one of k possible energies $E_1, E_2, \ldots, E_k$. A *configuration* of the gas is a distribution of the particles whereby n_1 of them have energy E_1, n_2 have energy $E_2, \ldots$ and n_k have energy E_k, where

$n_1 + n_2 + \cdots + n_k = N$. Since the order is irrelevant, we see that the total possible number of distinct configurations is

$$\binom{N}{n_1, n_2, \ldots, n_k} = \frac{N!}{n_1! n_2! \ldots! n_k!}.$$

Systems of particles whose configurations are counted in this way are said to be subject to *Boltzmann statistics* (a better terminology would be 'Boltzmann counting') in honour of Ludwig Boltzmann, a nineteenth-century physicist who was one of the founders of statistical mechanics. Two alternative methods of counting are used in quantum mechanics to count configurations for gases, all of whose particles consist either of bosons or fermions. These are associated with Bose–Einstein and Fermi–Dirac statistics respectively.

2.5 The gamma function (∗)

When we begin to learn mathematics, we encounter a number of important functions, which play a prominent role in both theory and applications, such as polynomials, exponentials, logarithms and the trigonometric functions. In this section, we will meet another function of this type whose role is to generalise $n!$ to all real numbers. The following lemma gives a major clue as to how to do this.

Lemma 2.6 *For all non-negative whole numbers n*

$$n! = \int_0^\infty e^{-x} x^n dx.$$

Proof We use induction. First observe that when $n = 0$, we have

$$\int_0^\infty e^{-x} dx = 1 = 0!$$

so the result is true in this case. Now suppose that the result holds for some n; then using integration by parts, we find that

$$\int_0^\infty e^{-x} x^{n+1} dx = [-e^{-x} x^{n+1}]_0^\infty + \int_0^\infty e^{-x}(n+1)x^n dx$$

$$= (n+1) \int_0^\infty e^{-x} x^n dx = (n+1)n!$$

$$= (n+1)!$$

where we have used the fact that for $n = 0, 1, 2, \ldots$, $\lim_{x \to \infty} e^{-x} x^n = 0$. $\qquad\square$

Based on Lemma 2.6, we define the *gamma function* $\Gamma(\alpha)$ for all real positive values α, by

$$\Gamma(\alpha) = \int_0^\infty e^{-x} x^{\alpha-1} dx$$

so that when α is a positive whole number, we have

$$\Gamma(\alpha) = (\alpha - 1)!.$$

(We use $x^{\alpha-1}$ rather than the more natural x^α in the integral for historical reasons.)

The gamma function extends the 'factorial' property to all positive real numbers as the following result shows.

Lemma 2.7 *For all positive real numbers α*

$$\Gamma(\alpha + 1) = \alpha \Gamma(\alpha).$$

Proof Integration by parts as in the previous lemma. $\quad\square$

We will meet the gamma function again in later chapters. It features in some intriguing formulae, and in Appendix 3 you may find the result

$$\Gamma(1/2) = \sqrt{\pi}$$

which seems a long way from $n!!$

Exercises

2.1. A game is constructed in such a way that a small ball can travel down any one of three possible paths. At the bottom of each path there are four traps which can hold the ball for a short time before propelling it back into the game. In how many alternatives ways can the game evolve thus far?

2.2. In how many different ways can seven coloured beads be arranged (a) on a straight wire, (b) on a circular necklace?

2.3. Prove Theorem 2.3.

2.4. Obtain $\binom{6}{2}$, $\binom{6}{3}$ and $\binom{7}{4}$ at first *without* using a calculator, then check your results on the machine.

2.5. In how many different ways (e.g. in a game of poker) can five distinct cards be dealt from a hand of 52?

2.6. Find n if $\binom{n}{8} = \binom{n}{4}$.

2.7. Show that $\binom{n}{r} + \binom{n}{r+1} = \binom{n+1}{r+1}$ and comment on the relationship between this identity and the form of Pascal's triangle.

2.8. In a lottery, the top three prizes, which are all different and of diminishing value, are allocated by drawing out of a hat. If there are 15 entrants, how many different winning combinations are there?

2.9. In a party game a blindfolded player has three turns to correctly identify one of the other seven people in the room. How many different possibilities are there for the resolution of the game?

2.10. A young couple go out to buy four grapefruits for a dinner party they are hosting. The man goes to greengrocer A where he has a choice of ten and the woman goes to greengrocer B where she can choose from eight. In how many different ways can the grapefruit be bought?

2.11. A firm has to choose seven people from its R and D team of ten to send to a conference on computer systems. How many ways are there of doing this

   (a) when there are no restrictions?
   (b) when two of the team are so indispensable that only one of them can be permitted to go?
   (c) when it is essential that a certain member of the team goes?

2.12. By putting $x = y = 1$ in the binomial theorem, show that

$$\sum_{r=0}^{n} \binom{n}{r} = 2^n.$$

2.13. Prove that the number of ways in which N objects can be divided into r groups of size $n_1, n_2, \ldots, n_r$ is

$$\binom{N}{n_1, n_2, \ldots, n_r}.$$

2.14. Show that

$$\binom{N}{n_1, n_2, \ldots, n_{k-1}, 0} = \binom{N}{n_1, n_2, \ldots, n_{k-1}}.$$

2.15. Calculate the number of ways in which out of a group of 12 balls, three can be put into a red box, five into a blue box and the rest into a yellow box.

2.16. Show that the number of ways that r indistinguishable objects can be placed into n different containers is

$$\binom{n + r - 1}{r}.$$

[*Hint*: Model the containers by adding $(n - 1)$ barriers to the r objects.]

2.17. Repeat problem (16) with the proviso that each container must contain no less than m objects (so $r > nm$). Show that when $m = 1$, the number of ways is

$$\binom{r - 1}{n - 1}.$$

2.18.* Define the Beta function by

$$\beta(m, n) = \int_0^1 x^{m-1}(1 - x)^{n-1} dx$$

where m and n are positive real numbers:

(i) Substitute $x = \cos^2(\theta)$ to show that

$$\beta(m, n) = 2 \int_0^{\pi/2} \cos^{2m-1}(\theta) \sin^{2n-1}(\theta) d\theta.$$

(ii) By substituting $y^2 = x$, show that

$$\Gamma(\alpha) = 2 \int_0^\infty e^{-y^2} y^{2\alpha-1} dy.$$

(iii) By use of polar coordinates and (ii) show that

$$\Gamma(m)\Gamma(n) = 2\Gamma(m+n) \int_0^{\pi/2} \cos^{2m-1}(\theta) \sin^{2n-1}(\theta) d\theta.$$

and hence deduce that

$$\beta(m, n) = \frac{\Gamma(m)\Gamma(n)}{\Gamma(m+n)}.$$

(iv) Show that when m and n are positive integers

$$\beta(m+1, n) = \left[n \binom{m+n}{m} \right]^{-1}.$$

Further reading

Most books on probability theory have a section on combinatorics which covers essentially the same material as given above. Two books are recommended in particular, not just for this chapter, but for all the subsequent ones on probability, these being *A First Course in Probability Theory* by Sheldon Ross (MacMillan, 1st edn 1976, 4th edn 1994) and *Elementary Probability Theory with Stochastic Processes* by K. L. Chung (Springer-Verlag, 1979). An interesting book that takes the application of combinatorics to probability theory far beyond the scope of the present volume is *Combinatorial Chance* by F. N. David and D. E. Barton (Charles Griffin and Co. Ltd, 1962). An account of Boltzmann statistics will be obtained in any elementary textbook on statistical mechanics. My own favourite is *Statistical Physics* by A. Isihara (Academic Press, 1971). The gamma function is the simplest of a large number of objects usually referred to as 'special functions' (the Beta function of Exercises 2.18) is another. A classic reference for these (and much more besides) is E. T. Whittaker and G. N. Watson, *A Course of Modern Analysis* (Cambridge University Press, 1920).

3
Sets and measures

3.1 The concept of a set

A *set* is simply a list of symbols. Usually, each symbol in the set is the name of an object, which may be conceptual in origin or which may have a physical existence. The *members* or *elements* of such a set are usually displayed within braces { } and separated by commas. For example, in later chapters we will be much concerned with dice games where the set of interest will contain the possible numbers which can appear on the faces of a dice. We will denote this set by S_1; then we have

$$S_1 = \{1, 2, 3, 4, 5, 6\}.$$

It is useful to have a notation to indicate when a particular element is a member of a certain set. We use the Greek letter $\in$ for this purpose; for example, it is clear that 2 is a member of the set S_1. We write this as

$$2 \in S_1$$

and say that 2 belongs to S_1.

Sometimes we may want to point out that a symbol of interest is not a member of some set. This is carried out using the symbol $\notin$, so that we have, for example.

$$7 \notin S_1.$$

The only example of a set that we have considered so far consists of numbers, but this is by no means typical. For example, the set S_2 consists of the names of the planets in the solar system:

$$S_2 = \{\text{Mercury, Venus, Earth, Mars, Jupiter, Saturn, Uranus,}$$
$$\text{Neptune, Pluto}\}.$$

Both of the sets we have considered so far are *finite*, that is they have only finitely many members; however, many important sets are *infinite*. An example is the set $\mathbb{N}$ of *natural numbers*:

$$\mathbb{N} = \{1, 2, 3, 4, \ldots\}.$$

The notation $\ldots$ indicates that the list continues *ad infinitum* and there is no 'finishing point'. The set $\mathbb{Z}$ of *integers* is infinite in both directions:

$$\mathbb{Z} = \{\ldots, -4, -3, -2, -1, 0, 1, 2, 3, 4, \ldots\}.$$

The fractions or *rational numbers* are usually denoted by $\mathbb{Q}$ and we have

$$\mathbb{Q} = \left\{ \cdots, -\frac{2}{3}, -4, -\frac{1}{3}, -3, -\frac{1}{2}, -2, -1, 0, 1, 2, \frac{1}{2}, 3, \frac{1}{3}, 4, \frac{2}{3}, \ldots \right\}.$$

Finally, we consider the set $\mathbb{R}$ of *real numbers*, that is all those numbers which represent distances from a fixed reference point on a straight line of infinite extent in both directions (Fig. 3.1). $\mathbb{R}$ has such a complex structure that we don't even try to list its elements as we have done for $\mathbb{Z}$ and $\mathbb{Q}$; however, we note that $\pi \in \mathbb{R}$, $\pi \notin \mathbb{Q}$, $\sqrt{2} \in \mathbb{R}$ and $\sqrt{2} \notin \mathbb{Q}$, for example.

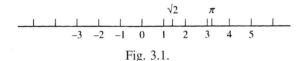

Fig. 3.1.

It is useful to have a notation for the number of elements in a set. For this we use the symbol #, so that we have, for example

$$\#(S_1) = 6, \quad \#(S_2) = 9, \quad \#(\mathbb{N}) = \#(\mathbb{Z}) = \#(\mathbb{Q}) = \infty, \quad \#(\mathbb{R}) = \infty^*.$$

Note: The reader may be puzzled that $\mathbb{N}$, $\mathbb{Z}$ and $\mathbb{Q}$ all have the same (infinite) number of elements, even though it appears that there are more numbers in $\mathbb{Q}$ than there are in $\mathbb{Z}$ and more in $\mathbb{Z}$ than there are in $\mathbb{N}$. Furthermore, I have denoted the number of elements in $\mathbb{R}$ by the unexplained symbol ∞^*, about which I am going to tell you nothing except that it is a bigger number than ∞! Further discussion of these 'paradoxes of the infinite' really goes beyond the scope of this book; however, a nice introduction to these mysteries can be found in Kasner and Newman (see references on page 40).

To bring us down to earth, we will introduce the notion of a *subset*. A set R is a subset of a set S if R comprises some but not necessarily all of the elements of S. If this is the case, we write

$$R \subseteq S.$$

For example:

if $R_1 = \{2, 4, 6\}$, we have $R_1 \subseteq S_1$;
if $R_2 = \{$Mercury, Venus, Earth, Mars$\}$, then $R_2 \subseteq S_2$.

The *singleton sets* of a given set are all those subsets comprising just a single element; for example, the singleton sets of S_1 are $\{1\}$, $\{2\}$, $\{3\}$, $\{4\}$, $\{5\}$ and $\{6\}$. It is important to distinguish between, for example, 1 and $\{1\}$: $1 \in S_1$ but $\{1\} \subseteq S_1$.

Note: Some books like to distinguish between proper and improper subsets. Thus R is a *proper subset* of S if R is a subset and we exclude the possibility that R might be S itself. In this case we write $R \subset S$. In all the examples above we could have written $\subset$ instead of $\subseteq$. Nonetheless, in the rest of the book we will always use $\subseteq$, for when we consider subsets in probability theory, we do not want to exclude the possibility that R might indeed turn out to be S itself.

Subsets of $\mathbb{R}$ are usually defined by means of some recipe which must be satisfied for all members. Important examples are the *intervals*: let a and b be two real numbers with $a < b$. The *closed interval* $[a, b]$ is the subset of $\mathbb{R}$ comprising all those real numbers which lie between a and b, as shown in Fig. 3.2.

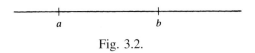

Fig. 3.2.

A formal definition of $[a, b]$ is

$$[a, b] = \{x \in \mathbb{R}; a \leq x \leq b\}.$$

Note that in the definition of $[a, b]$ the semi-colon; should be read as 'such that'.

We observe that each of the end-points a and b are themselves members of $[a, b]$. For example

$$0 \in [0, 1], 0.73 \in [0, 1], 1.0001 \notin [0, 1], [0.5, 0.75] \subseteq [0, 1].$$

It will be useful for us to also have the concept of an *open interval* where the end-points are not included. These are usually represented using round brackets so

$$(a, b) = \{x \in \mathbb{R}; a < x < b\}.$$

For example

$$0 \notin (0, 1), 1 \notin (0, 1), \ldots .$$

Intermediate between the closed and open intervals are the *half-open intervals* where only one end-point is included. These are defined by

$$(a, b] = \{x \in \mathbb{R}, a < x \leq b\}.$$

and

$$[a, b) = \{x \in \mathbb{R}, a \leq x < b\}.$$

For example

$$0 \notin (0, 1], 1 \in (0, 1], 0 \in [0, 1), 1 \notin [0, 1).$$

Our concept of 'interval' includes those that extend off to infinity in one or both directions, for example $(-\infty, a]$, $[b, \infty)$ and $(-\infty, \infty) = \mathbb{R}$.

Note that when dealing with intervals which are infinite in extent we always use a round bracket next to the infinity to indicate that we never actually reach an infinite point no matter how hard we strive. This is a technical point which you should not worry about too much.

We often find ourselves dealing with a large number of subsets of a given set. In this situation the *Venn diagram* is a useful device for giving us a quick insight into the relationships between the various subsets. Suppose, for example, that R_1, R_2, R_3 and R_4 are all subsets of S, and that we also know that $R_2 \subseteq R_1$, R_1 and R_3 share some elements in common but R_4 has no elements in common with any of the other three subsets. We would represent this situation in the Venn diagram shown in Fig. 3.3. The set S, which is drawn as a rectangular box, is sometimes called the *universal set* when all other sets under consideration are subsets of it.

Exercise: Find examples of subsets $R_1, \ldots, R_4$ satisfying the above conditions when S is (a) S_1 (dice), (b) S_2 (planets).

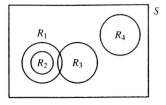

Fig. 3.3.

3.2 Set operations

In this section we will discover methods of making new sets from old. For this purpose we will consider a universal set S with subsets $A, B, C, \ldots$ throughout.

The first set we construct is the *complement* $\overline{A}$ of the set A which is simply the set of all elements of S which do not belong to A. Symbolically, we have

$$\overline{A} = \{x \in S; x \notin A\}.$$

$\overline{A}$ is represented as the shaded part of the Venn diagram in Fig. 3.4. For example, if S is S_1 and $A = \{1, 3, 5\}$, then $\overline{A} = \{2, 4, 6\}$.

Sets and measures

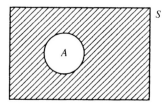

Fig. 3.4.

Now for two sets A and B, their *union* $A \cup B$ is the set of all elements of S which either belong to A or belong to B or belong to both A and B. (So $\cup$ represents the 'inclusive–or' concept.) The precise definition is

$$A \cup B = \{x \in S; x \in A \text{ or } x \in B \text{ or } x \in \text{ both } A \text{ and } B\}.$$

The Venn diagram in Fig. 3.5 shows $A \cup B$ shaded. For example, if $S = S_2$ and (denoting planets by their initial letters) $A = \{\text{Me, E, J, N}\}$, $B = \{\text{E, Ma, J, S}\}$, then $A \cup B = \{\text{Me, E, Ma, J, S, N}\}$.

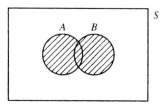

Fig. 3.5.

Finally, we introduce the *intersection* of the two sets A and B, which we denote as $A \cap B$. These are precisely those elements of S which are members of both A and B. Symbolically, we have

$$A \cap B = \{x \in S; x \in A \text{ and } x \in B\}$$

and this is shown pictorially in Fig. 3.6. For example, with $S = S_2$ and A, B as above, we have $A \cap B = \{E, J\}$. Note that

$$A \cap B \subseteq A \subseteq A \cup B$$

and

$$A \cap B \subseteq B \subseteq A \cup B$$

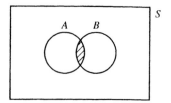

Fig. 3.6.

which you can check by drawing a Venn diagram.

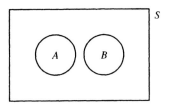

Fig. 3.7.

The *empty set* Ø is defined to be that set which has no elements whatsoever; Ø may seem a somewhat redundant object but we will find it serves many useful purposes. Formally, we have

$$\emptyset = \overline{S};$$

Ø is a subset of every set S. This may seem strange initially; to see that it is true first of all observe that every subset of S can be obtained by removing various combinations of its elements. Then Ø is the subset we get when we take away *all* the elements of S.

The subsets A and B are said to be *disjoint* if they have no elements in common, that is if

$$A \cap B = \emptyset.$$

Pictorially, the Venn diagram in Fig. 3.7 demonstrates disjoint sets.

Exercise: Find pairs of disjoint subsets in both S_1 and S_2.

In the study of logic we aim to deduce the truth or falsity of propositions from that of other, sometimes simpler, propositions. A key technique for making new propositions from old is the use of the logical connectives 'not', 'or' and 'and'. We have the following analogy with set theory:

SET THEORY	LOGIC
Subset	Proposition
Complement	Not
Union	Or (inclusive)
Intersection	And

Sometimes we have to simplify quite complicated expressions involving many strings of subsets. The *laws of Boolean algebra* facilitate these procedures (Table 3.1). Properties (B2)–(B5) can all be checked by the reader by drawing suitable Venn diagrams to compare the left- and right-hand sides of each expression. Items (B1), (B6) and (B7) are self-evident.

Table 3.1. *Boolean laws*

(B1)	Idempotency	$A \cup A = A$; $A \cap A = A$
(B2)	Associativity	$A \cap (B \cap C) = (A \cap B) \cap C$ $A \cup (B \cup C) = (A \cup B) \cup C$
(B3)	Commutativity	$A \cap B = B \cap A$; $A \cup B = B \cup A$
(B4)	Distributivity	$A \cap (B \cup C) = (A \cap B) \cup (A \cap C)$ $A \cup (B \cap C) = (A \cup B) \cap (A \cup C)$
(B5)	de Morgan's laws	$\overline{A \cup B} = \overline{A} \cap \overline{B}$ $\overline{A \cap B} = \overline{A} \cup \overline{B}$
(B6)	Properties of the complement	$\overline{\overline{A}} = A$; $A \cap \overline{A} = \emptyset$ $A \cup \overline{A} = S$
(B7)	Properties of S and $\emptyset$	$A \cup S = S$; $A \cap S = A$ $A \cup \emptyset = A$; $A \cap \emptyset = \emptyset$

Note that associativity (B2) allows us to give an unambiguous meaning to such expressions as $A \cup B \cup C$ and $A \cap B \cap C$, for example

$$A \cup B \cup C = (A \cup B) \cup C = A \cup (B \cup C).$$

A simple example of the value of the Boolean laws in practice follows.

Example 3.1 Simplify the expression

$$A \cap \overline{(A \cap \overline{B})}.$$

Solution

$$A \cap \overline{(A \cap \overline{B})} = A \cap (\overline{A} \cup \overline{\overline{B}}) \qquad \text{by (B5)}$$

$$= A \cap (\overline{A} \cup B) \qquad \text{by (B6)}$$

$$= (A \cap \overline{A}) \cup (A \cap B) \qquad \text{by (B4)}$$

$$= \emptyset \cup (A \cap B) \qquad \text{by (B6)}$$

$$= A \cap B. \qquad \text{by (B7)}$$

Note: $^{-}$, $\cup$ and $\cap$ are operations which make new sets from old. There are many other such operations, some of which we will meet below. Mathematicians like to pare discussions down to the smallest number of fundamental operations from which all the others can be built. In this regard we could take $^{-}$ and $\cup$ (or alternatively $^{-}$ and $\cap$) and then *define* $\cap$ (or alternatively $\cup$) by means of de Morgan's law (B5)

$$A \cap B = \overline{\overline{A} \cup \overline{B}}$$

An auxiliary operation we will find very useful is the *difference* of two sets $A - B$, which is defined to be all those members of A that are not members of B. The difference can be expressed in terms of $^-$ and $\cap$ by

$$A - B = A \cap \overline{B}.$$

The Venn diagram for $A - B$ is shown in Fig. 3.8. Observe that if A and B are disjoint then $A - B = A$.

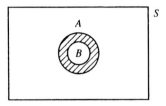

Fig. 3.8.

3.3 Boolean algebras

As a set can be a list of any 'things' we like, there is no good reason why we can't have sets whose elements are themselves sets; for example, consider the set S_1 discussed above and form the set

$$T_1 = \{\{1, 1\},\ \{1, 2\},\ \{1, 3\},\ \{1, 4\},\ \{1, 5\},\ \{1, 6\}\}.$$

T_1 may seem a strange object but it can be given an interpretation (if you throw two die in succession, T_1 describes all possible outcomes in which one of the die is always a 1).

One of the most important sets of sets is the *power set* of a set. The power set of S is the set whose elements are all the subsets of S, and we will denote it by $\mathcal{P}(S)$. Clearly, we must have $\emptyset \in \mathcal{P}(S)$ and $S \in \mathcal{P}(S)$. For example, if

$$S = \{a, b, c\}$$

then

$$\mathcal{P}(S) = \{\emptyset, \{a\}, \{b\}, \{c\}, \{a, b\}, \{b, c\}, \{a, c\}, S\}.$$

Note that in the example given above, $\#(S) = 3$ and $\#\mathcal{P}(S) = 8 = 2^3$. The following result will be very useful in the sequel.

Theorem 3.1 *If $\#(S) = n$, then $\#\mathcal{P}(S) = 2^n$.*

Proof We begin to count the elements of $\mathcal{P}(S)$, starting with those subsets with the fewest members: $\emptyset$ counts for 1, there are n singleton sets, the number of subsets of S with two members is precisely $\binom{n}{2}$, and continuing in this manner we obtain

$$\#(\mathcal{P}(S)) = 1 + n + \binom{n}{2} + \binom{n}{3} + \cdots + n + 1$$

$$= (1+1)^n = 2^n. \qquad \text{by Exercises 2.12}$$

$\square$

Quite often we will be interested in a subset of $\mathcal{P}(S)$, that is a set containing some, but not necessarily all, of the subsets of S, for example $T_1 \subseteq \mathcal{P}(S_1)$. Let $\mathcal{B}(S) \subseteq \mathcal{P}(S)$; we say that $\mathcal{B}(S)$ is a *Boolean algebra* if:

(i) whenever $A \in \mathcal{B}(S)$, then $\overline{A} \in \mathcal{B}(S)$,
(ii) whenever $A \in \mathcal{B}(S)$ and $B \in \mathcal{B}(S)$, then $A \cup B \in \mathcal{B}(S)$.

Notice that if $\mathcal{B}(S)$ is a Boolean algebra, it follows immediately from (i) and (ii) that if $A \in \mathcal{B}(S)$ and $B \in \mathcal{B}(S)$, then $A \cap B \in \mathcal{B}(S)$ as $A \cap B = \overline{\overline{A} \cup \overline{B}}$. As $\mathcal{B}(S) \subseteq \mathcal{P}(S)$, $\mathcal{B}(S)$ must contain at least one set; in fact both S and $\emptyset$ belong to every Boolean algebra, for if $A \in \mathcal{B}(S)$, then by (i) $\overline{A} \in \mathcal{B}(S)$, hence by (ii) $A \cup \overline{A} = S \in \mathcal{B}(S)$ and by (i) again $\emptyset = \overline{S} \in \mathcal{B}(S)$.

Clearly, T_1 is not a Boolean algebra as both (i) and (ii) are violated. We now give some examples of subsets of $\mathcal{P}(S)$ which are Boolean algebras. Obviously, $\mathcal{P}(S)$ itself is a Boolean algebra.

Example 3.2

$$\mathcal{L}(S) = \{\emptyset, S\}.$$

You may check that $\mathcal{L}(S)$ is a Boolean algebra by using rule (B7) and the definition of $\emptyset$, noting that, by (B6), we have $\overline{\emptyset} = S$. Although $\mathcal{L}(S)$ may seem a rather uninteresting example, it is in fact the basic tool underlying the logic used in building electronic circuits. The flow of electricity through such circuits is controlled by switches, which may be either on or off. Each such switch is represented by a Boolean variable x, which takes the value S when the switch is on and $\emptyset$ when the switch is off. Two switches x and y in series are represented by the intersection $x \cap y$ and in parallel by the union $x \cup y$. The complement $\overline{x}$ describes a switch which is on when x is off and off when x is on. More details can be found in many engineering books but note that engineers often write $x \cap y$ as xy and $x \cup y$ as $x + y$ (see Fig. 3.9).

Example 3.3 Let $a \in S$, then the 'Boolean algebra generated by a' is

$$\mathcal{B}_a(S) = \{\emptyset, \{a\}, S - \{a\}, S\}.$$

You should convince yourself that $\mathcal{B}_a(S)$ is the smallest Boolean algebra of subsets of S that contains $\{a\}$.

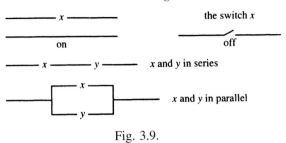

Fig. 3.9.

The following example will be very important in our subsequent development of continuous probability. Recall the set of real numbers $\mathbb{R}$ and the notion of an open interval (a, b).

Example 3.4 $\mathcal{I}(\mathbb{R})$ is defined to be the smallest Boolean algebra containing all open intervals (a, b), where $-\infty \leq a < b \leq \infty$.

To obtain an understanding of what the most general kind of set in $\mathcal{I}(\mathbb{R})$ is, we observe that

$$\overline{(a, b)} = -(\infty, a] \cup [b, \infty) \in \mathcal{I}(\mathbb{R}).$$

Also, each closed interval $[a, b] \in \mathcal{I}(\mathbb{R})$ since

$$[a, b] = \overline{(-\infty, a)} \cap \overline{(b, \infty)}.$$

Singleton sets $\{a\}$ comprising isolated points are also in $\mathcal{I}(\mathbb{R})$ since

$$\{a\} = [a, \infty) - (a, \infty).$$

Now we can write down the most general member of $\mathcal{I}(\mathbb{R})$. Suppose that $I_1, I_2, \ldots, I_n$ are all distinct mutually disjoint sets where each of the I_js is either an interval or an isolated point. We call the union

$$J = I_1 \cup I_2 \cup \cdots \cup I_n \text{ a } broken \text{ } line \text{ in } \mathbb{R}.$$

Fig. 3.10 shows a broken line when $n = 4$.

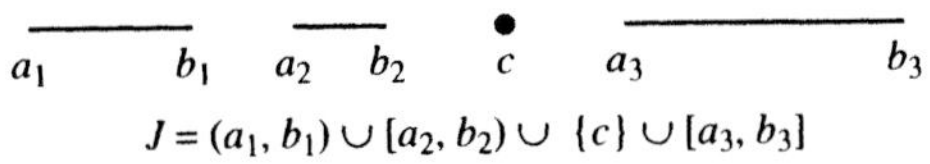

$$a_1 \qquad b_1 \quad a_2 \quad b_2 \quad c \quad a_3 \qquad b_3$$
$$J = (a_1, b_1) \cup [a_2, b_2) \cup \{c\} \cup [a_3, b_3]$$

Fig. 3.10.

You should convince yourself that the complement of a broken line is again a broken line and that the union of two broken lines is also a broken line, so that $\mathcal{I}(\mathbb{R})$ really is a Boolean algebra (see Exercise 3.8 for a practical example to help you see this). Of course, each interval is itself a broken line (the case $n = 1$) and we will regard $\emptyset$ as a broken line containing no intervals (the case $n = 0$).

On some occasions we will want to work with the broken lines, which lie wholly within an interval $[a, b]$. In this case the relevant Boolean algebra will be written $\mathcal{I}(a, b)$.

3.4 Measures on Boolean algebras

In this section we introduce the important concept of a measure or weight of a set. Probability as defined in the next chapter will be a special case of this notion.

Let S be a set and $\mathcal{B}(S)$ a Boolean algebra of subsets of S. A (finite) *measure m* is the assignment of a non-negative real number $m(A)$ to every set $A \in \mathcal{B}(A)$ such that the following condition (M) is satisfied.

If A and $B \in \mathcal{B}(S)$ and they are disjoint (i.e. $A \cap B = \emptyset$), then

$$m(A \cup B) = m(A) + m(B). \tag{M}$$

Notes

 (i) Technically m is a mapping (or function) from $\mathcal{B}(S)$ to $[0, \infty]$.
 (ii) Most textbooks give a more refined definition of measure than that described above. Instead of a Boolean algebra they use a more complex object called a σ-algebra and the defining condition (M) is replaced by a stronger condition called σ-additivity. These refinements allow for a much more satisfactory treatment of convergence within measure theory. However, in an introductory book like this one they can be omitted without any great harm – but if you want to know more, see the note at the end of the chapter.

If $A, B, C \in \mathcal{B}(S)$ are mutually disjoint, we find

$$m(A \cup B \cup C) = m(A \cup (B \cup C)) = m(A) + m(B \cup C) \qquad \text{by (M)}$$
$$= m(A) + m(B) + m(C) \qquad \text{by (M) again.}$$

You can extend this result to an arbitrary number of mutually disjoint subsets in Exercise 3.9(iii).

When we measure properties of physical objects such as mass or weight we find the quantity of interest to be distributed over its form in various proportions. The measure concept generalises this notion to arbitrary sets.

Example 3.5 Let S be a finite set and take the Boolean algebra to be $\mathcal{P}(S)$. For each $A \in \mathcal{P}(S)$, define

$$m(A) = \#(A)$$

then A is a measure called the *counting measure*. To verify (M) we simply observe that if A and B are disjoint and $\#(A) = m$, $\#(B) = n$, then $\#(A \cup B) = m + n$.

Example 3.6 Let a and b be positive real numbers with $a < b$. Take $S = [a, b]$ and $\mathcal{B}(S) = \mathcal{I}(a, b)$ and define

$$m([c, d]) = d - c \quad \text{for } [c, d] \subseteq [a, b]$$

then it is easy to see that m is a measure. In fact, m just gives the length of the interval $[c, d]$. The measure m is called the *Lebesgue measure*, after the French mathematician Henri Lebesgue, who used it as the basis of a very powerful theory of integration which he developed at the beginning of the twentieth century.

Example 3.7 In the same set-up as Example 3.6, let f be a positive function (i.e $f(x) \geq 0$ for all $a \leq x \leq b$) which satisfies $\int_a^b f(x)dx < \infty$ and define

$$m([c, d]) = \int_c^d f(x)dx \quad \text{for } [c, d] \subseteq [a, b].$$

To see that (M) is satisfied, recall that the definite integral evaluates the area under the curve f in Fig. 3.11. In fact, if $[c_1, d_1]$ and $[c_2, d_2]$ are disjoint intervals in $[a, b]$, we can define

$$\int_{[c_1, d_1] \cup [c_2, d_2]} f(x)dx = \int_{c_1}^{d_1} f(x)dx + \int_{c_2}^{d_2} f(x)dx$$

which is natural from the area point of view (see Fig. 3.12).

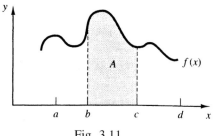

Fig. 3.11.

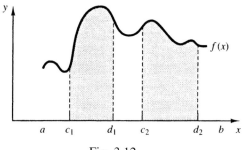

Fig. 3.12.

The reader may be familiar from calculus courses with a special case of the above where the points d_1 and c_2 coincide so that $[c_1, d_1] \cup [c_2, d_2] = [c_1, d_2]$ and we have

$$\int_{c_1}^{d_2} f(x) dx = \int_{c_1}^{d_1} f(x) dx + \int_{d_1}^{d_2} f(x) dx.$$

We will meet many measures of this type in the chapter on continuous random variables. This example generalises the previous one in that we recapture the Lebesgue measure when we put $f(x) = 1$ for all $a \leq x \leq b$.

Note that in both Examples 3.6 and 3.7 we have $m(\{a\}) = 0$ for isolated points so that whenever $a < b$, by Exercise 3.9(iii)

$$m([a,b]) = m(\{a\} \cup (a, b) \cup \{b\})$$
$$= m(\{a\}) + m((a, b)) + m(\{b\}) = m((a, b)).$$

The reader should be warned that this does not hold for all measures on $\mathcal{I}(a, b)$.

In Examples 3.6 and 3.7, we have defined measures directly on intervals rather than broken lines which are the most general members of $\mathcal{B}(S)$. However, if $J = I_1 \cup I_2 \cup \cdots \cup I_n$ is a broken line, then by Exercise 3.9(iii) we have

$$m(J) = m(I_1) + m(I_2) + \cdots + m(I_n)$$

so that the measures of all broken lines can be calculated once we know the prescription on intervals; for example, if m is the Lebesgue measure on $[0, 10]$ and $J = [1, 3] \cup \{5\} \cup [7, 10]$, then

$$m(J) = m([1, 3]) + m(\{5\}) + m([7, 10])$$
$$= 2 + 0 + 3 = 5.$$

Note: If $J = [a_1, b_1] \cup [a_2, b_2] \cup \cdots \cup [a_n, b_n]$, then we define

$$\int_J f(x) dx = \int_{a_1}^{b_1} f(x) dx + \int_{a_2}^{b_2} f(x) dx + \cdots + \int_{a_n}^{b_n} f(x) dx.$$

We will find this notation useful in Chapter 8.

Example 3.8 Let $S = \mathbb{R}$ and $\mathcal{B}(S) = \mathcal{I}(\mathbb{R})$. Fix a number $a \in \mathbb{R}$. We define a measure which is usually denoted δ_a by

$$\delta_a(J) = 1 \qquad \text{if } a \in J,$$
$$\delta_a(J) = 0 \qquad \text{if } a \notin J$$

for any broken line J. δ_a is called the *Dirac measure at a* after the great British physicist Paul Dirac, who invented it for his research on quantum mechanics in the mid-twentieth century. It is now widely used throughout mathematics and its applications.

Examples 3.5–3.8 should have convinced you that measure is a wide-ranging concept and in the next chapter it will form the basis of our approach to probability. The following result will be invaluable for us.

Theorem 3.2 *Let m be a measure on* $\mathcal{B}(S)$ *for some set S. Let* $A, B \in \mathcal{B}(S)$:

(i) *If* $B \subseteq A$, *then*
$$m(A - B) = m(A) - m(B).$$

(ii) *If* $B \subseteq A$, *then*
$$m(B) \leq m(A).$$

(iii) $m(\emptyset) = 0.$
(iv) *If A and B are arbitrary, then*
$$m(A \cup B) = m(A) + m(B) - m(A \cap B).$$

Proof

(i) Check on a Venn diagram that
$$(A - B) \cup B = A \text{ and } (A - B) \cap B = \emptyset$$
thence by (M)
$$m(A) = m(A - B) + m(B)$$
and the result follows.
(ii) Immediate from (i) since $m(A - B) \geq 0$.
(iii) Since $\emptyset = S - S$ we have by (i)
$$m(\emptyset) = m(S) - m(S) = 0.$$

(iv) Check on a Venn diagram that
$$A \cup B = (A - (A \cap B)) \cup (B - (A \cap B)) \cup (A \cap B);$$
then using the result of Exercises 3.9(iii) and (i) we obtain
$$m(A \cup B) = m(A - (A \cap B)) + m(B - (A \cap B)) + m(A \cap B)$$
$$= m(A) - m(A \cap B) + m(B) - m(A \cap B) + m(A \cap B)$$
$$= m(A) + m(B) - m(A \cap B), \text{ as required.}$$

$\square$

Note that (iv) generalises (M) to arbitrary (i.e. not necessarily disjoint) A and B. By (ii) it follows that the maximum value of $m(A)$ for any set A is $m(S)$. We call $m(S)$ the *total mass* of the measure m. For example, in Example 3.6 given above, if $(a, b) = (0, 1)$ and m is the Lebesgue measure on $\mathcal{I}(0, 1)$, then it has total mass 1.

If we now take $(a, b) = (5, 9)$ and m to be the Lebesgue measure on $\mathcal{I}(5, 9)$, then it has total mass 4.

Finally, we will need the concept of a *partition* of a set S. This is a family $\mathcal{E} = \{E_1, E_2, \ldots, E_n\}$ with each $E_j \subseteq S (1 \leq j \leq n)$ such that:

(i) the E_js are mutually disjoint, that is

$$E_j \cap E_k = \emptyset \qquad \text{for all } 1 \leq j, k \leq n \text{ with } j \neq k.$$

(ii) $E_1 \cup E_2 \cup \cdots \cup E_n = S$.

The Venn diagram in Fig. 3.13 illustrates a partition with $n = 5$. Every set can be partitioned into its singleton sets, for example

$$S_1 = \{1\} \cup \{2\} \cup \{3\} \cup \{4\} \cup \{5\} \cup \{6\}.$$

The following result can easily be deduced from (ii) above and Exercise 3.9(iii), and we will find it of great value in the next chapter.

Let S be a finite set and $\mathcal{E}$ a partition of S with each $E_j \in \mathcal{B}(S)$; then if m is a measure on $\mathcal{B}(S)$, we have

$$m(S) = \sum_{j=1}^{n} m(E_j).$$

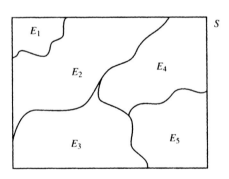

Fig. 3.13.

Note: Measures on σ-algebras (*)

Let $\sigma(S) \subseteq \mathcal{P}(S)$. We say that $\sigma(S)$ is a σ-*algebra* if:

(i) whenever $A \in \mathcal{B}(S)$, then $\overline{A} \in \mathcal{B}(S)$,
(ii) whenever the sequence $A_1, A_2, \ldots \in \mathcal{B}(S)$, then

$$\bigcup_{n=1}^{\infty} A_n \in \mathcal{B}(S).$$

Essentially in a σ-algebra, the property of Boolean algebras given in Exercise 3.9(i) is extended to include infinite unions. A measure m on $\sigma(S)$ is then a map from $\sigma(S)$ to $[0, \infty)$ such that whenever the sequence $A_1, A_2, \ldots \in \sigma(S)$ and each $A_i \cap A_j = \emptyset (1 \leq i, j < \infty)$, then

$$m \left(\bigcup_{n=1}^{\infty} A_n \right) = \sum_{n=1}^{\infty} m(A_n)$$

so that Exercise 3.9(iii) is extended to infinite series. If you want to do probability theory completely rigorously, then you should use measures on σ-algebras rather than on Boolean algebras. I have chosen to do the latter in this book because the differences are mainly technical and the Boolean algebra approach makes life a lot easier for you. However, there are places in this book where I have cheated a little bit and am really using σ-algebras without saying so. Can you find these?

Exercises

3.1. Consider the following subsets of $S_1 = \{1, 2, 3, 4, 5, 6\}$: $R_1 = \{1, 2, 5\}$, $R_2 = \{3, 4, 5, 6\}$, $R_3 = \{2, 4, 6\}$, $R_4 = \{1, 3, 6\}$, $R_5 = \{1, 2, 5\}$. Find:

(a) $R_1 \cup R_2$,
(b) $R_4 \cap R_5$,
(c) $\overline{R_5}$,
(d) $(R_1 \cup R_2) \cap R_3$,
(e) $\overline{R_1} \cup (R_4 \cap R_5)$,
(f) $\overline{(R_1 \cap (R_2 \cup R_3))}$,
(g) $(R_1 \cup \overline{R_2}) \cap (R_4 \cup \overline{R_5})$.

3.2. Express the following sets in $\mathbb{R}$ as a single interval:

(a) $\overline{(-\infty, 1) \cup (4, \infty)}$,
(b) $[0, 1] \cap [0.5, 2]$,
(c) $[-1, 0] \cup [0, 1]$.

3.3. By drawing a suitable Venn diagram, convince yourself of the following 'laws of absorption'

$$A \cap (A \cup B) = A$$
$$A \cup (A \cap B) = A.$$

3.4. Deduce the following from the Boolean laws:

(i) $S \cup S = S$,
(ii) $S \cap S = S$,
(iii) $\emptyset \cup \emptyset = \emptyset$,
(iv) $\emptyset \cap \emptyset = \emptyset$,

(v) $S \cup \emptyset = S$,
(vi) $S \cap \emptyset = \emptyset$.

3.5. Use the laws of Boolean algebra to establish the following properties of the set theoretic difference:

(i) $A \cap (B - C) = (A \cap B) - (A \cap C)$,
(ii) $(A \cup B) - C = (A - C) \cup (B - C)$,
(iii) $A - (A - B) = A \cap B$,
(iv) $(A - B) - C = (A - C) - B = A - (B \cup C)$,
(v) $A - (A \cap B) = A - B$.

3.6. We've already seen that $\cup$ is analogous to the logical 'inclusive or'. The 'exclusive or' (which is used more commonly in everyday speech) is analogous to the operation $\odot$ defined by

$$A \odot B = (A \cup B) - (A \cap B)$$

so $A \odot B$ means either A or B but not both.

(a) Use the Boolean laws to deduce that:

(i) $A \odot B = B \odot A$,
(ii) $A \odot \emptyset = A$,
(iii) $A \odot A = \emptyset$,
(iv) $A \odot S = \overline{A}$,
(v) * $A \cap (B \odot C) = (A \cap B) \odot (A \cap C)$.

(b) Verify that

$$A \odot B = (A - B) \cup (B - A).$$

3.7. Suppose two coins are tossed successively with each coin being a head (H) or a tail (T). If the outcome is a head on the first coin and a tail on the second coin, we write this outcome as HT, etc. So the set of all possible outcomes (S_3) is

$$S_3 = \{HH, HT, TH, TT\}.$$

Write $\mathcal{P}(S_3)$.

3.8. If $J_1 = (-\infty, -9] \cup [-5, -2] \cup [0, 4] \cup [6, 9]$ and $J_2 = [-49, -23] \cup [-2, 1] \cup [9, 15]$ are two broken lines in $\mathcal{I}(\mathbb{R})$, write an expression for $J_1 \cup J_2$ and confirm that it is a broken line.

3.9. Let S be a set and M a measure on $\mathcal{B}(S)$. If $A_1, A_2, \ldots, A_n \in \mathcal{B}(S)$ are all mutually disjoint, use the technique of mathematical induction to prove that:

(i) $A_1 \cup A_2 \cup \cdots \cup A_n \in \mathcal{B}(S)$,
(ii) $A_1 \cap A_2 \cap \cdots \cap A_n \in \mathcal{B}(S)$,
(iii) $m(A_1 \cup A_2 \cup \cdots \cup A_n) = \sum_{j=1}^{n} m(A_j)$.

3.10. Show that if A, B and C are arbitrary (not necessarily disjoint) sets in $\mathcal{B}(S)$, then

$$m(A \cup B \cup C) = m(A) + m(B) + m(C) - m(A \cap B)$$
$$-m(A \cap C) - m(B \cap C) + m(A \cap B \cap C).$$

[*Hint*: Write $A \cup B \cup C = A \cup (B \cup C)$ and apply Exercise 3.9(iii) via Theorem 3.2(iv).]

3.11. Let S be a set with total mass 1 and $\mathcal{P}$ a partition into sets E_1, E_2, E_3 and E_4. It is known that $m(E_1) = 0.5$, $m(E_2) = 0.25$ and $m(E_3) = 0.125$. Find $m(E_4)$.

3.12. Let f be a function on $[a,b]$. The volume of revolution generated by rotating f about the y-axis is

$$V([a,b]) = \pi \int_a^b f(x)^2 \mathrm{d}x.$$

Verify that V is a measure on $\mathcal{I}(a, b)$.

3.13. In Example 3.7 of measures given by definite integrals, compute the following:

(a) $m([0, 1])$, when $f(x) = x^3$,
(b) $m([2, \infty))$, when $f(x) = e^{-x}$,
(c) $m([1, 2] \cup [3, 4])$, when $f(x) = 4 - x$,
(d) $m(J)$, where J is the broken line $(1, 4] \cup \{5\} \cup (8, 10) \cup \{12\}$ and $f(x) = \frac{1}{x}$.

3.14. Show that if m and n are two measures on $\mathcal{B}(S)$ and $c \geq 0$, then:

(i) $m + n$ is a measure on $\mathcal{B}(S)$, where $(m + n)(A) = m(A) + n(A)$,
(ii) cm is a measure, where $(cm)(A) = c \cdot m(A)$.

What are the total masses of these measures?

3.15. Let S be a finite set and α a positive function on S so that $\alpha(x) \geq 0$ for each $x \in S$. For each $A \in \mathcal{P}(S)$, define

$$m(A) = \sum_{x \in A} \alpha(x)$$

where the sum is over all those elements x which lie in A. Convince yourself that m is a measure on $\mathcal{P}(S)$. [*Note*: These measures are the discrete analogues of those described in Example 3.7.]

3.16. Let $\mathcal{P} = \{E_1, E_2, \ldots, E_n\}$ be a partition of S and $A \in \mathcal{B}(S)$. By means of a Venn diagram, or otherwise, show that $\mathcal{P}_A = \{E_1 \cap A, E_2 \cap A, \ldots, E_n \cap A\}$ is a partition of A. Hence, show that if P is a measure on $\mathcal{B}(S)$, then

$$P(A) = \sum_{i=1}^n P(A \cap E_i).$$

3.17. Let m be a measure on $\mathcal{B}(S)$, where S is a finite set. Define the 'distance' $d(A, B)$ between two sets A and $B \in \mathcal{B}(S)$ by

$$d(A, B) = m(A \odot B).$$

Show that d satisfies the triangle inequality

$$d(A, C) \leq d(A, B) + d(B, C).$$

[*Hint*: Show that $d(A, B) + d(B, C) - d(A, C) \geq 0$.]

3.18.* Let S be a finite set. If $B, A \subseteq S$, we define the relative complement of B in A, $A \backslash B$ by

$$A \backslash B = A - B \qquad \text{if } B \subseteq A$$

$$= \emptyset \text{ otherwise.}$$

Now let m and n be two measures on $\mathcal{P}(S)$ and define the convolution of m and n, written $m * n$, by

$$(m * n)(A) = \sum_{B \subseteq S} m(A \backslash B) n(B)$$

where $A \in \mathcal{P}(S)$ and the sum is over all subsets B of S. Show that:

  (i) $m * n$ is a measure on $\mathcal{P}(S)$,

  (ii) $m * n = n * m$,

  (iii) if p is another measure on $\mathcal{P}(S)$, then $(m * n) * p = m * (n * p)$.

Further reading

Set theory is the language which underlies all modern abstract mathematics. In fact, at one time it was believed that all mathematics could be founded upon it. There are many fine books on the subject but for a sound introduction with the minimum of fuss the classic *Naive Set Theory* by Paul Halmos (Springer-Verlag, 1974) is unbeatable. For a gentle (i.e. bedtime reading) introduction to the paradoxes of the infinite and to much else besides read *Mathematics and the Imagination* by E. Kasner and J. Newman (Jarrold and Sons Ltd, 1949). Boolean algebra is named after the British mathematician George Boole, whose fundamental work *The Laws of Thought* (Dover, 1958), first published in 1854, is quite readable. The origins of the Boolean laws can be seen in Chapter 2 of this. Modern mathematicians see Boolean algebra as a special case of the study of more fundamental objects called lattices. For information about these and many of the other beautiful structures of modern algebra see *Algebra* by Saunders Maclane and Garrett Birkhoff (Macmillan, 1971).

Measure theory is one of the cornerstones of twentieth-century mathematics. The reader should be warned that it is a heavily technical subject which should not be tackled without first gaining a sound background in real analysis. Again, there are many books – in particular, the reader might like to look at Paul Halmos' *Measure Theory* (Springer-Verlag, 1974) or a volume with the same title by Donald L. Cohn (Birkhaüser, 1993). If you want to learn both real analysis and measure theory together, then H. L. Royden's *Real Analysis* (3rd edn, Collier Macmillan, 1988) will give you a sound foundation for advanced study.

4
Probability

4.1 The concept of probability

Suppose we are given a set S, a Boolean algebra $\mathcal{B}(S)$ and a measure P on $\mathcal{B}(S)$. We say that P is a *probability* (or *probability measure*) if it has total mass 1. In the discussion below, we will attempt to justify this definition. First, though, it may be worthwhile summarising some of the more useful properties of probabilities, most of which we have already met in Chapter 3 in the section on measures.

4.1.1 Properties of probabilities

Let A and $B \in \mathcal{B}(S)$, then we have:

(P1) $0 \le P(A) \le 1$.
(P2) $P(S) = 1$, $P(\emptyset) = 0$.
(P3) $P(A \cup B) = P(A) + P(B)$ if A and B are disjoint.
(P4) $P(A - B) = P(A) - P(B)$ if $B \subseteq A$.
(P5) $P(\overline{B}) = 1 - P(B)$.
(P6) $P(A) \le P(B)$ whenever $A \subseteq B$.
(P7) $P(A \cup B) = P(A) + P(B) - P(A \cap B)$ for arbitrary A and B.
(P8) If $\mathcal{E} = \{E_1, E_2, \ldots, E_n\}$ is a partition of S, then $\sum_{j=1}^{n} P(E_j) = 1$.

These properties are all either part of the definition of (probability) measure or are results from Theorem 3.2. The only unfamiliar one may be (P5), which is easily deduced from (P4) and (P2) by putting $A = S$ and using the fact that $S - B = \overline{B}$. We note that (P1) and (P3), together with the condition $P(S) = 1$, may be taken as axioms for the concept of probability (see Section 1.3). In probability theory, the abstract set-up described above and in the preceding chapter is given a practical interpretation, which allows us to use it to model chance phenomena. This interpretation is as follows.

Suppose we want to model some experience which we know is going to have an unpredictable outcome such as rolling a fair die. The elements of the set S are all the possible outcomes of the experience. For example, for the die game, we know that one of the six faces must show, so

$$S = S_1 = \{1, 2, 3, 4, 5, 6\}.$$

Convention: In probability theory, S is called the *sample space* and its elements are called *outcomes*.

Now suppose that we want to place a bet on some feature of the experience actually being realised. It is natural to ask the question: 'What is the most general type of bet that can be made?'

Clearly, we can bet on one of the outcomes arising, but we can also make more complicated bets. For example, in S_1 we can bet on an even number occurring, that is on one of the members of the set $\{2, 4, 6\}$, or we can bet on a number greater than 2 occurring, that is we can bet on a member of the set $\{3, 4, 5, 6\}$. A short consideration should convince you that you can bet on any subset of S_1, that is any member of $\mathcal{P}(S_1)$, which is, of course, a Boolean algebra.

In general, $\mathcal{P}(S)$ may be too large for our purposes; however, the most general type of bet that can be made can always be described as one of the subsets of S belonging to a suitable Boolean algebra $\mathcal{B}(S)$.

Convention: In probability theory, the subsets in $\mathcal{B}(S)$ are called *events*. An event A is realised whenever the outcome of the experience is a member of A. You should convince yourself that the set theoretic operations of union, intersection and complement can be interpreted as 'inclusive-or', 'and' and 'not', respectively, when applied to events.

Note: In probability theory, provided the set S is finite or has $\#(S) = \infty$, we will always take $\mathcal{B}(S) = \mathcal{P}(S)$. This type of model is called *discrete probability* and in this and the next chapter we will mainly concentrate on this case. When S is an interval $[a, b]$ in $\mathbb{R}$, we will take $\mathcal{B}(S) = \mathcal{I}(a, b)$. We are then in the domain of *continuous probability*, which we will deal with extensively in Chapters 8 and 9 below.

Now the role of P is to assign a weight between 0 and 1 to each event in $\mathcal{B}(S)$. We interpret a weight of 1 to mean *certainty* and 0 to mean *impossibility*, so that the interpretation of (P2) is that, given that S contains all possible outcomes, it is certain that when the experience happens one of them must occur and it is impossible that none of them occurs. If $P(A)$ is close to 1, for example $P(A) = 0.95$, then A is nearly certain and we would be very surprised if the outcome of the experience is not a member of A. On the other hand, if $P(A)$ is close to 0, for example $P(A) = 0.07$, then A is very unlikely and we would again express surprise if the outcome of the experience is a member of A. Events A for which $P(A)$ is close to 0.5 have greater unpredictability and, indeed, if $P(A) = 0.5$, we say that A has *maximum uncertainty*. (The notion of uncertainty will be made more precise in Chapter 6.)

Note: The triple comprising the sample space S, the Boolean algebra $\mathcal{B}(S)$ of events and the probability measure P is often written $(S, \mathcal{B}(S), P)$ and called a *probability space*.

4.2 Probability in practice

So far, our notion of probability is a fairly abstract thing and, apart from the terminology we are using, seems to have little bearing on the real world. In this section we will attempt to clarify this relationship. We begin by contemplating some of the ways in which chance enters our lives. Consider the following statements:

(S1) There is a one in two chance that this coin will come up tails.
(S2) There's a strong chance that Wales will beat Denmark in the match tonight.
(S3) Of 16 seeds planted in identical conditions, nine germinated successfully so the chances of a successful germination are 0.56.

Each of these statements manifests a different way in which chance is used in everyday language. We will examine each of them in turn.

4.2.1 Probability by symmetry

Consider a fair coin which we know to have no defects. Apart from the fact that one side is a head and the other a tail, there are no other features which distinguish one side from the other. In this situation we say that we have a *symmetry* between the two outcomes. When the coin is tossed, Nature has no means of favouring one side over the other, so it seems reasonable to make the following assignment

$$S = \{H, T\}, \mathcal{B}(S) = \mathcal{P}(S), P(H) = 0.5, P(T) = 0.5$$

where we have written $P(\{H\}) = P(H)$ for convenience.

In general, we will make the following assumption.

Principle of symmetry

Let S be a finite sample space with outcomes $\{a_1, a_2, \ldots, a_n\}$, which are such that all of the a_js are physically identical except for the labels we attach to them. In this case we always take

$$P(a_1) = P(a_2) = \cdots = P(a_n)$$

unless we are given evidence to the contrary.

Another example where the principle of symmetry applies is to the dice throwing game where the sample space is S_1.

Now suppose the principle of symmetry applies to a set S where $\#(S) = n$, and let p denote the common value of all the $P(a_j)$s. Let $\mathcal{P}$ be the partition of S into its singleton sets; then by (P8) we have

$$\sum_{j=1}^{n} p = 1, \text{ i.e. } np = 1, \text{ i.e. } p = \frac{1}{n}.$$

Furthermore, if A is any event in S and $\#(A) = r$ (say), where $r \le n$, then a similar argument to the above gives

$$P(A) = \frac{r}{n}.$$

These results make the principle of symmetry very easy to apply in practice.

Example 4.1 A fair die is thrown. Find the probability of (a) throwing a 6, (b) throwing an even number, (c) not throwing a 5, (d) throwing a number less than 3 or more than 5.

Solution

(a) We have $S = S_1$. By the principle of symmetry, as $\#(S_1) = 6$ we obtain $P(6) = \frac{1}{6}$.
(b) The event is $A = \{2, 4, 6\}$. Clearly, $\#(A) = 3$, so $P(A) = \frac{3}{6} = \frac{1}{2}$.
(c) Let B be the event that a 5 is thrown so $P(B) = \frac{1}{6}$. We want $P(\bar{B}) = 1 - P(B) = \frac{5}{6}$ (by (P5)).
(d) Let A be the event that a number less than 3 is thrown so that $A = \{1, 2\}$ and let B be the event that a number greater than 5 is thrown so that $B = \{6\}$; then we require $P(A \cup B)$ and since A and B are disjoint we use (P3) to obtain

$$P(A \cup B) = P(A) + P(B) = \frac{2}{6} + \frac{1}{6} = \frac{1}{2}$$

Of course, the probabilities in (c) and (d) could have been worked out directly without using (P3) and (P5), but it is useful to gain experience with these properties as they will be indispensible in more complicated problems.

The principle of symmetry has a wider range of application than may at first appear to be the case. In many situations, such as quality inspections of manufactured products or collecting samples of rainwater for analysis of pollutants, we have a large set of items which are not necessarily identical in every way but we treat them as if they are identical so that they all have an equal chance of being chosen. This procedure is called *random sampling* and is an important tool in statistics.

Example 4.2 Thirty manufactured items have been inspected and it is known that five are defective. If three items are chosen at random, what is the probability that all of them are defective?

Solution The sample space S for this problem consists of all possible selections of three items from the 30. It would be extremely tedious to list all of these but fortunately we don't have to. All we will need is $\#(S) = \binom{30}{3} = 4060$. The event A in this problem consists of those selections of three items in which all are defective. Again we only need $\#(A) = \binom{5}{3} = 10$.

So, by the principle of symmetry

$$P(A) = \frac{10}{4060} = 0.002\,46.$$

Note that in this problem the key words 'chosen at random' alert us to the possibility of using the principle of symmetry.

Subjective probabilities

Let us return to the statement (S2). How strong a chance is there that Wales will beat Denmark in the match? Clearly, there is no objective criterion we can use to assign such a probability. Different people are likely to offer different opinions based on their own preconceived ideas, particularly supporters of Wales and Denmark. Nonetheless, we can still incorporate such ideas into our scheme.

Let a denote the proposition 'Wales will beat Denmark in tonight's match' and $\bar{a}$ be its negation 'Wales will not beat Denmark in tonight's match'. If we are concerned solely with these propositions, we work with the set $S = \{a, \bar{a}\}$ and the relevant Boolean algebra is $\mathcal{B}_a(S) = \{\emptyset, \{a\}, \{\bar{a}\}, S\}$ (see Example 3.3). If I then believe that the probability that Wales will beat Denmark is 0.8, I assign $P(a) = 0.8$ so that $P(\bar{a}) = 0.2$ and we have a perfectly well-defined probability (we have again condensed $P(\{a\})$ to $P(a)$ in our notation).

Subjective (or personal) probabilities are often expressed in terms of *odds*, and this gives us a good opportunity to investigate their relationship with probabilities. In fact, this is quite simple – suppose that a bookmaker is offering odds of x to y on Wales beating Denmark. This means that out of a total of $(x + y)$ equally valued coins, s(h)e is willing to bet x of them that Wales beat Denmark; in other words, his/her personal probability is

$$P(a) = \frac{x}{x + y}.$$

In bookmakers' language, odds of x to y against a are the same as odds of x to y on $\bar{a}$.

Example 4.3 A bookmaker is offering odds of 7–2 against Arctic Circle winning the 3.15 at Haydock Park. What is the probability that s(h)e is assigning to Arctic Circle winning the race?

Solution Let *a* be the proposition that Arctic Circle wins the race, then with $x = 7$ and $y = 2$ we have

$$P(\overline{a}) = \frac{7}{9}$$

thus, by (P5)

$$P(a) = 1 - \frac{7}{9} = \frac{2}{9}.$$

Subjective probability may appear to be a rather intangible concept; however, it can be given a more precise (operational) meaning as follows.

Let us return to proposition *a* representing Wales beating Denmark in the match. Suppose that you, the reader, and I, the author, are going to have a bet about the result of this match based on me offering you certain odds. Suppose that I offer odds of 5:1 on *a*; this means that if Wales loses, I must pay you £5, but if Wales wins, you only have to give me £1. Based on the above recipe, my personal probability is $P(a) = 0.833$. This seems to represent a pretty good bet for you, but you may think that you could get better terms so you hesitate and, as a consequence, I offer you better odds of 6:1 so that $P(a)$ is now 0.857.

Further attempts to haggle are fruitless, and it is clear that the maximum odds I will offer you are 6:1. We then say that my personal probability $P(a)$ is established as 0.857. This discussion leads us to offer the following definition of personal probability $P(a)$ (sometimes called the *degree of belief* in *a*).

$P(a)$ is that probability which is determined by the maximum odds I will offer on *a*.

Note that subjective probabilities can also be applied to individuals' beliefs about experiences that have already occurred, for example many people may be unsure, if asked, about the precise year in which the American president J. F. Kennedy died. For example, person A might be prepared to offer odds of 2 to 1 that it was 1962 (it was in fact 1963).

We will continue our discussion of subjective probability in Section 4.5 after we have met the concept of conditional probability.

Relative frequency

In our discussion above, we have developed a theory of probability that can be applied to very general situations, namely experiences with indeterminate outcomes. A more traditional approach to probability would regard this as too vague and insist that we only speak of probability in the context of strictly controlled *experiments*, which can be repeated under the same conditions as many times as we like. An example of such a set-up occurs in statement (S3) above. Here the experiment is the germination of seeds and we require each seed to be planted in identical conditions as regards soil quality, room temperature and humidity, etc.

In (S3), nine of the 16 seeds germinated. In a more general set-up, we can imagine n identical experiments being carried out. If we are monitoring a particular outcome x and we observe that x occurs in r of the n experiments, we define the *relative frequency* of x based on n experiments (which we denote as $f_n(x)$) by

$$f_n(x) = \frac{r}{n}$$

so that in (S3) above, x is a successful germination, $n = 16$, $r = 9$ and $f_{16}(x) = 0.56$.

Although relative frequencies can be interpreted as probability measures, there are obvious practical problems in doing so. For example, in (S3), after carrying out our 16 germination experiments, we may be inclined to say that the probability of germination $P(x) = 0.56$. However, suppose we now try another experiment under the same conditions. If the experiment results in a germination, we have $P(x) = f_{17}(x) = \frac{10}{17} = 0.588$ and if it fails, then $P(x) = f_{17}(x) = \frac{9}{17} = 0.529$. Either way, we see that we are in the undesirable situation where $P(x)$ depends on the number of experiments carried out and is not an absolute property of seedlings.

A popular approach to solving this problem, which is described in many statistics textbooks, is to use a mathematical device that you may have met before called 'taking the limit' and define

$$P(x) = \lim_{n \to \infty} f_n(x).$$

The idea behind this is that we will get a true picture of the regularity of the germination of seeds if we can look at an infinite number of identical experiments. Of course, we cannot carry out an infinite number of experiments in practice but, the argument goes, if we can do a 'large enough number', we will be able to see the trend.

A serious problem with this whole approach is that there is no guarantee that the limit actually exists in any given experimental arrangement. If you have studied an elementary analysis course, you will know that a sequence defined in terms of a known formula can either converge, diverge or oscillate, and limits only exist for convergent series; for example, the sequence $(1/n)$ converges and has limit 0 but the sequence $(1 - (-1)^n)$ oscillates between the two values 0 and 2. In the sequences which are generated by relative frequencies, we have no nice formula to which we can apply the known mathematical tests for convergence; indeed, we do not know what the nth term will be until we have carried out the nth experiment. Consequently, there is no mathematical basis for the definition of probabilities as limits of relative frequencies. From what we have seen above, we cannot interpret relative frequencies as probabilities. However, unless we can find some way of incorporating relative frequencies into our theory, we will be in a great deal of trouble, as a probabilistic interpretation of relative frequencies is essential for statistics. We will return to this problem in Section 4.5.

Exercise: Toss a coin 100 times and regard each separate toss as a repetition of the experiment – tossing a coin. Let x be the outcome 'heads'. Write down $f_{10}(x)$, $f_{20}(x)$, ..., $f_{100}(x)$. See if you can decide at any stage whether or not your coin is biased. Write down your personal probability $P(x)$ after each run of ten experiments.

4.3 Conditional probability

Suppose that you are going to buy a second-hand car. You find one that you like and after a brief inspection you convince yourself that it would have a high probability of being roadworthy. Being a cautious individual, you then inspect the car again in the company of your friend, who is a skilled mechanic and who discovers a large number of major problems. Consequently, you decrease your personal probability of the car being roadworthy.

In the above example we have seen an important general principle at work: namely:

> *Gain in information about an event leads to a change in its probability.*

To formalise this idea suppose we have a value for the probability of an event A, $P(A)$. Now suppose that we obtain the information that another event B has occurred. In the light of our knowledge of B, the probability of A now changes to $P_B(A)$, which we call the *conditional probability of A given B*. Our next goal is to obtain a method of calculating $P_B(A)$.

Note: Most textbooks write $P_B(A)$ as $P(A/B)$. We will not use this latter notation (at least not in the main part of this book) as it misleadingly suggests that conditional probability is the probability of some set A/B constructed from A and B.[†]

Example 4.4 Consider Table 4.1 describing babies born in one week at a certain hospital. If a baby is chosen at random and turns out to be a boy, find the probability that it was born to a mother over the age of 30 years.

Solution The sample space S for this problem consists of all the babies born in the hospital in the week in question. Note that $\#(S) = 168$. Now let A be the event that a baby is born to a mother who is more than 30 years old and let B be the event that a baby is a boy.

As we know that the baby we've selected is a boy, we will only be interested in the top row of the table, and from that row it is the left-hand entry that gives the

---

[†] As we'll see in Chapter 10, the notation $P(A|B)$ can be useful when B is a complicated set containing many outcomes.

Table 4.1.

Number of babies born which are	Born to mothers over 30 years old	Born to mothers less than 30 years old
Boys	34	49
Girls	41	44

number of boys born to mothers over 30 years. So, by the principle of symmetry, the probability we want is

$$P_B(A) = \frac{34}{83} = \frac{\#(A \cap B)}{\#(B)}.$$

To write $P_B(A)$ directly in terms of a ratio of probabilities, note that

$$P_B(A) = \frac{\#(A \cap B)}{\#(S)} \div \frac{\#(B)}{\#(S)}$$

$$= \frac{P(A \cap B)}{P(B)}.$$

In the last example, our commonsense led us to a very useful formula for conditional probabilities. We will now adopt this as a formal definition.

Given a probability space $(S, \mathcal{B}(S), P)$ and events $A, B \in \mathcal{B}(S)$ where $P(B) \neq 0$, the *conditional probability* of A given B, $P_B(A)$ is defined by

$$P_B(A) = \frac{P(A \cap B)}{P(B)}.$$

Note that in many situations $P_A(B)$ may make just as much sense as $P_B(A)$, for example in Example 4.4 above $P_A(B)$ is the probability that a randomly chosen baby that is born to a woman over the age of 30 years turns out to be a boy and you should check that in this case $P_A(B) = \frac{34}{75}$. We will explore the relationship between $P_B(A)$ and $P_A(B)$ in greater detail below. First we will develop some of the properties of conditional probability.

In the following we will always assume that $B \in \mathcal{P}(S)$ has $P(B) \neq 0$.

Theorem 4.1

(a) Let A_1 and A_2 be disjoint subsets in $\mathcal{B}(S)$, then

$$P_B(A_1 \cup A_2) = P_B(A_1) + P_B(A_2).$$

(b) If $A \in \mathcal{B}(S)$, then

$$P_B(\overline{A}) = 1 - P_B(A).$$

Proof

(a) First note that if A_1 and A_2 are disjoint, then so are $A_1 \cap B$ and $A_2 \cap B$, as can be confirmed by drawing a Venn diagram.

Now

$$P_B(A_1 \cup A_2) = \frac{P((A_1 \cup A_2) \cap B)}{P(B)}$$

$$= \frac{P((A_1 \cap B) \cup (A_2 \cap B))}{P(B)} \qquad \text{by (B4)}$$

$$= \frac{P(A_1 \cap B) + P(A_2 \cap B)}{P(B)} \qquad \text{by (P3)}$$

$$= P_B(A_1) + P_B(A_2).$$

(b)

$$P_B(\overline{A}) = \frac{P(\overline{A} \cap B)}{P(B)} = \frac{P(B - A)}{P(B)}$$

$$= \frac{P(B - A \cap B)}{P(B)} \qquad \text{by Exercises 3.5(v)}$$

$$= \frac{P(B) - P(A \cap B)}{P(B)} \qquad \text{by (P4)}$$

$$= 1 - P_B(A).$$

□

The results of Theorem 4.1 suggest that conditional probabilities behave similarly to probability measures. We will explore this further in the exercises.

The definition of conditional probability tells us that

$$P(A \cap B) = P_B(A)P(B). \qquad (4.1)$$

It is sometimes useful to extend this to three or more events, for example

$$P(A \cap B \cap C) = P((A \cap B) \cap C) \qquad \text{by (B2)}$$

$$= P_{A \cap B}(C)P(A \cap B)$$

$$= P_{A \cap B}(C)P_A(B)P(A). \qquad (4.2)$$

Conditional probability arguments sometimes give an alternative to combinatorial approaches to problems.

Example 4.5 Use conditional probability to solve the problem of Example 4.2. Find also the probability that of two items selected at random, one is defective.

Solution Let A, B and C be the events that the first, second and third items selected are defective (respectively). We require $P(A \cap B \cap C)$. Using the principle of symmetry, we find that

$$P(A) = \frac{5}{30}, \; P_A(B) = \frac{4}{29} \quad \text{and} \quad P_{A \cap B}(C) = \frac{3}{28}$$

so, by (4.2)

$$P(A \cap B \cap C) = \frac{5}{30} \times \frac{4}{29} \times \frac{3}{28} = 0.002\,46.$$

For the second problem we require $P((A \cap \overline{B}) \cup (\overline{A} \cap B))$. As the two events in brackets are disjoint, we obtain

$$P((A \cap \overline{B}) \cup (\overline{A} \cap B)) = P(A \cap \overline{B}) + P(\overline{A} \cap B) \qquad \text{by (P3)}$$

$$= P(A)P_A(\overline{B}) + P(\overline{A})P_{\overline{A}}(B)$$

$$= \frac{5}{30} \times \frac{25}{29} + \frac{25}{30} \times \frac{5}{29} = 0.287$$

where we have used Theorem 4.1(b).

A further useful result on conditional probabilities is the following.

Theorem 4.2 *Let* $\mathcal{P} = \{E_1, E_2, \ldots, E_n\}$ *be a partition of S with each $P(E_i) \neq 0$ and let $A \in \mathcal{P}(S)$, then*

$$P(A) = \sum_{i=1}^{n} P_{E_i}(A)P(E_i).$$

Proof You should not find it difficult to verify that the sets $A \cap E_1, A \cap E_2, \ldots, A \cap E_n$ yield a partition of A, so that

$$A = (A \cap E_1) \cup (A \cap E_2) \cup \cdots \cup (A \cap E_n).$$

Hence, by Exercise 3.9(iii)

$$P(A) = \sum_{i=1}^{n} P(A \cap E_j)$$

$$= \sum_{i=1}^{n} P_{E_i}(A)P(E_i) \text{by}(4.1).$$

$\square$

Application (a simple communication system)

Here we begin to make contact with the second main theme of this book, namely the communication of information across a noisy channel which distorts the incoming signal. We will consider the simplest possible model, called the *binary symmetric channel*, which is sketched in Fig. 4.1.

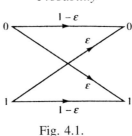

Fig. 4.1.

The input consists of just the two symbols 0 and 1 (which could represent the terms 'off' and 'on' or 'no' and 'yes'). Signals are sent across the channel, where they are picked up by a receiver.

We write $P(1)$ for the probability that a 1 is sent and $P(0)$ for the probability that a 0 is sent. Since these are the only possibilities, we have

$$P(0) + P(1) = 1.$$

For simplicity, we write $P(1) = p$ so that $P(0) = 1 - p$. The noise in the channel distorts the signal so that the probabilities of receiving a 0 or a 1 as output are changed. Let $Q(0)$ and $Q(1)$ be the probabilities that a 0 or a 1 are received. As above, we write $Q(1) = q$ so that $Q(0) = 1 - q$.

Finally, we describe the effect of the noise which distorts the signal. (This could be a fault in the circuits if the channel is electrical, or static caused by bad weather for radio reception.) The effect of the noise is captured perfectly by introducing conditional probabilities, so that $P_0(1)$ is the probability that a 1 is received given that a 0 is transmitted and $P_1(0)$ is the probability that a 0 is received given that a 1 is transmitted. We also have the probabilities of successful transmission $P_0(0)$ and $P_1(1)$.

By Theorem 4.1 (a) we have

$$P_0(1) + P_0(0) = 1$$

and

$$P_1(0) + P_1(1) = 1.$$

In the binary *symmetric* channel which we are considering here, we always take $P_0(1) = P_1(0) = \varepsilon$ so that $P_1(1) = P_0(0) = 1 - \varepsilon$.

Suppose that you are the receiver and that you know the values of p and ε. We can now calculate q by using Theorem 4.2

$$\begin{aligned}
Q(1) &= P_0(1)P(0) + P_1(1)P(1) \\
&= \varepsilon(1 - p) + (1 - \varepsilon)p \\
&= \varepsilon + p - 2\varepsilon p
\end{aligned}$$

and

$$Q(0) = P_1(0)P(1) + P_0(0)P(0)$$
$$= \varepsilon p + (1 - \varepsilon)(1 - p)$$
$$= 1 - \varepsilon - p + 2\varepsilon p.$$

Clearly, $Q(0) + Q(1) = 1$, as was required.

In practice, we would like to keep ε as small as possible. Notice that if we put $\varepsilon = 0$ into the above formulae, we obtain $q = p$, as we would expect. Also, if the input probabilities are equal (i.e. $p = \frac{1}{2}$), we see immediately that the output probabilities are also equal (i.e. $q = \frac{1}{2}$), irrespective of the value of ε (see Exercise 4.15 below).

You may be wondering how to express the above example in terms of our usual language of sample spaces and events. This will be explained in Chapter 7.

Earlier, I promised to tell you more about the relationship between $P_A(B)$ and $P_B(A)$. This is the subject of the following result, usually known as *Bayes' theorem*.

Theorem 4.3

(a) *Let A and $B \in \mathcal{B}(S)$ with $P(A)$, $P(B) \neq 0$, then*

$$P_A(B) = \frac{P_B(A)P(B)}{P(A)}.$$

(b) *Let $\mathcal{P} = \{E_1, E_2, \ldots, E_n\}$ be a partition of S with each $P(E_i) \neq 0$ and let A and B be as above, then*

$$P_A(B) = \frac{P_B(A)P(B)}{\sum_{i=1}^{n} P_{E_i}(A)P(E_i)}.$$

Proof

(a) We have

$$P_A(B) = \frac{P(A \cap B)}{P(A)}$$
$$= \frac{P_B(A)P(B)}{P(A)} \quad \text{by (4.1)}.$$

(b) Obtained immediately by rewriting the denominator of the result in (a) by using the result of Theorem 4.2.

$\square$

Bayes' theorem seems like a fairly innocuous piece of mathematics; however, its use in statistics has aroused a great deal of controversy throughout this century. Statisticians who adopt an approach to their subject based on the use of Bayes' theorem are called 'Bayesians'. Bayesians tend also to be fervent believers in the use of subjective probabilities, and we will have more to say about this at the end of the chapter.

Example 4.6 We return to the situation of the binary symmetric channel discussed above. When the receiver obtains the message, which in this case is either a 0 or a 1, his/her main concern is with the reliability of the channel, that is with questions of the type: 'What is the probability that a 1 was sent out, given that a 1 has been received?'

Solution We write $Q_a(b)$ for the conditional probability that b was sent out given that a is received (where both a and b can be 0 or 1). As we found above for the $P_a(b)$s, we obtain

$$Q_0(0) + Q_0(1) = 1 \text{ and } Q_1(0) + Q_1(1) = 1.$$

Now by Bayes' theorem, we obtain

$$Q_0(0) = \frac{P_0(0)P(0)}{Q(0)} = \frac{(1-\varepsilon)(1-p)}{1-\varepsilon-p+2\varepsilon p}$$

and

$$Q_1(1) = \frac{P_1(1)P(1)}{Q(1)} = \frac{(1-\varepsilon)p}{\varepsilon+p+2\varepsilon p}.$$

We leave the calculation of $Q_0(1)$ and $Q_1(0)$ as an exercise for the reader.

We close this section by proving a 'conditional' version of Theorem 4.2, which we will find of use in Section 4.5 below.

Theorem 4.4 *Let A and $\mathcal{P}$ be as in the statement of Theorem 4.2, and let H be an event with $P(H) \neq 0$, then*

$$P_H(A) = \sum_{i=1}^{n} P_{H \cap E_i}(A) P_H(E_i).$$

Proof

$$P_H(A) = \frac{P(H \cap A)}{P(H)}$$

$$= \sum_{i=1}^{n} \frac{P(A \cap H \cap E_i)}{P(H)} \qquad \text{by Exercise 3.16}$$

$$= \sum_{i=1}^{n} \frac{P_{H \cap E_i}(A)P(H \cap E_i)}{P(H)} \qquad \text{by (4.1)}$$

$$= \sum_{i=1}^{n} P_{H \cap E_i}(A) P_H(E_i).$$

$\square$

4.4 Independence

In ordinary language, we use the word 'independence' to mean that two experiences are completely separate and the occurrence of one has no effect on the other. In probability theory, we use the word to mean that each experience has no effect on the *probability* which is assigned to the other. As Exercise 4.19 demonstrates, these two uses of the word should not be confused.

To make formal sense of the probabilistic term, let A and B be two events with $P(A) \neq 0$. Our notion of independence tells us that if A occurs, then $P(B)$ is unaltered (and vice versa). Since the occurrence of A leads to a gain in information so that $P(B)$ changes to $P_A(B)$, we say that A and B are *(probabilistically) independent* if

$$P_A(B) = P(B),$$

that is

$$\frac{P(A \cap B)}{P(A)} = P(B),$$

that is

$$P(A \cap B) = P(A)P(B). \tag{4.3}$$

Note that if $P(B) \neq 0$, it follows from (4.3) that $P_B(A) = P(A)$. In future, we will take (4.3) as the definition of probabilistic independence since it has the advantage of remaining valid when $P(A)$ or $P(B)$ is 0.

Note: Some books refer to 'probabilistic independence' as 'stochastic independence' or 'statistical independence'. We will often just use the term 'independence' where the context is obvious.

Property (P7) takes a very simple form when A and B are independent, which we will find useful later on

$$P(A \cup B) = P(A) + P(B) - P(A)P(B). \tag{4.4}$$

In our earlier discussion of relative frequencies, we discussed experiments which were repeated under identical conditions. It is usual to regard these as probabilistically independent (see, however, the discussion in the next section).

Example 4.7 A fair die is thrown twice in succession. What is the probability of a 6 on each occasion?

Solution Let A be the event that a 6 is thrown on the first occasion and B the event that a 6 is thrown on the second occasion.

By the principle of symmetry, we have

$$P(A) = P(B) = \frac{1}{6}.$$

Hence, by independence, we obtain

$$P(A \cap B) = P(A)P(B) = \frac{1}{36}.$$

The reader might like to compare this method with that of writing out the sample space for the combined throws of the die and then using the principle of symmetry directly.

The following result is very useful and accords well with our intuition:

Theorem 4.5 *If $P(A) \neq 0$ and A and B are probabilistically independent, then so are A and $\overline{B}$.*

Proof

$$
\begin{aligned}
P(A \cap \overline{B}) &= P_A(\overline{B})P(A) & \text{by (4.1)} \\
&= (1 - P_A(B))P(A) & \text{by Theorem 4.1(b)} \\
&= (1 - P(B))P(A) & \text{by independence} \\
&= P(A)P(\overline{B}) & \text{by (P5).}
\end{aligned}
$$

$\square$

A similar argument establishes that $\overline{A}$ and $\overline{B}$ are independent.

Example 4.8 A computer sales manager has a supply of 100 machines of a certain type. It is discovered that 17 of these have a problem with the hard disc and that 9 have faulty monitors. Given that these two problems are independent, find the probability that a machine chosen at random has:

(a) a hard disc problem only,
(b) neither problem.

Solution Let H be the event that a machine with a faulty hard disc is chosen and let M be the event that one with a bad monitor is selected. By the principle of symmetry and (P5), we obtain

$$P(H) = 0.17, \quad P(\overline{H}) = 0.83,$$
$$P(M) = 0.09, \quad P(\overline{M}) = 0.91.$$

(a) We require $P(H \cap \overline{M}) = P(H)P(\overline{M}) = 0.15$ by Theorem 4.5.
(b) $P(\overline{H} \cap \overline{M}) = P(\overline{H})P(\overline{M}) = 0.755.$

We have to be careful about extending the definition of independence to three or more events. In fact if A, B and C are events, we say that they are probabilistically independent if

$$P(A \cap B) = P(A)P(B), \ P(B \cap C) = P(B)P(C)$$

$$P(A \cap C) = P(A)P(C) \quad \text{and} \quad P(A \cap B \cap C) = P(A)P(B)P(C).$$

As is shown in Exercise 4.20, the last of these relations is not necessarily satisfied if you only assume that the first three are. More generally, we say that the events $A_1, A_2, \ldots, A_n$ are probabilistically independent if for every combination $i_1, i_2, \ldots, i_r$ of r events from the n we have

$$P(A_{i_1} \cap A_{i_2} \cap \cdots \cap A_{i_r}) = P(A_{i_1})P(A_{i_2})\ldots P(A_{i_r}).$$

You should check by means of a combinatorial argument that this involves satisfying $2^n - n - 1$ separate conditions.

[*Hint*: Use $$1 + n + \sum_{r=2}^{n} \binom{n}{r} = 2^n \qquad \text{(see Exercise 2.12).]}$$

4.5 The interpretation of probability

In Section 4.2 above we met three different ways in which probability is used in practice. Each of these has, at one time or another, been hailed as an all-encompassing approach to the whole of probability, and we will now look at each of them again in this light.

4.5.1 The classical theory of probability

This goes back to the founders of probability theory (see the next section for more information on this). The idea is to attempt to interpret all probabilities as arising from the principle of symmetry. This idea is not as crazy as it looks. It rests on the belief that if we analyse a problem carefully enough, we will eventually find that it can be broken up into pieces to which the principle of symmetry applies. An immediate objection to this would be that we cannot model bias using symmetry. However, consider the following set-up. A coin is inserted into a machine. After entering the slot it can go down either of two pathways. If it goes down the left pathway, the coin is kept by the machine but if it goes down the right pathway, we obtain a prize. The mechanism is symmetric, so, by the principle of symmetry, we have Fig. 4.2, with W representing 'win' and L representing 'lose'

$$P(W) = P(L) = \frac{1}{2}$$

Now suppose the manufacturers bias the machine so that the odds against winning are 2:1. It now appears that we can no longer use the principle of symmetry to obtain

Fig. 4.2.

the probabilities; however, we can model this set-up by introducing what a physicist
would call the 'internal states' X, Y and Z. When the coin enters the machine, the
arrangement of the paths X, Y and Z are symmetric so that, by the principle of
symmetry

$$P(X) = P(Y) = P(Z) = \frac{1}{3}$$

but the paths X and Y both result in the coin being kept, while only path Z allows
us to win the prize; hence (Fig. 4.3)

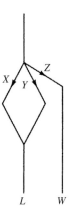

Fig. 4.3.

$$P(W) = P(Z) = \frac{1}{3} \text{ and } P(L) = P(X \cup Y) = P(X) + P(Y) = \frac{2}{3}.$$

It is not unreasonable to suggest that other instances of bias (e.g. in coins and die)
might be analysable in such a fashion. The founders of probability tried to extend
the 'principle of symmetry' to a more general statement called the 'principle of
insufficient reason'. This states that in any situation where you want to assign
probabilities, you should always make these equal unless you have some reason
to do otherwise. The principle is vague, and attempts to tidy it up and fashion it
into a reliable tool have not been successful. Another problem with it is that it
leads to a number of paradoxes. For example, suppose that a friend is knitting
you a new jumper and s(h)e won't tell you what colour it is. The principle of
insufficient reason encourages you to assign $P(\text{Red}) = \frac{1}{2}$ and $P(\text{Not red}) = \frac{1}{2}$.

It also advises you to assign $P(\text{Red}) = \frac{1}{4}$, $P(\text{Blue}) = \frac{1}{4}$, $P(\text{Green}) = \frac{1}{4}$ and $P(\text{Not red or green or blue}) = \frac{1}{4}$.[†]

Despite its defects there are situations where the 'principle of sufficient reason' can be used to great effect, as the following interesting example shows.

Example 4.9 A container contains two balls, each of which may be black or white. Twice in succession, a ball is removed from the container and then returned to it. On each occasion the colour of the ball is found to be white. If a ball is taken out of the container a third time, what is the probability that it, too, will be white?

Solution Let W_3 be the event that the ball is white on the third draw and $2W$ be the event that the ball is white on the first two drawings. So we want to calculate $P_{2W}(W_3)$.

At the beginning of the process there are three beliefs about the balls which it is possible to hold about their colour:

H_1: Both of the balls are white.
H_2: One of the balls is white and the other is black.
H_3: Both of the balls are black.

Since there is no reason to prefer any of these over the others, we use the 'principle of insufficient reason' to assign

$$P(H_1) = P(H_2) = P(H_3) = \frac{1}{3}.$$

Now, since the two drawings are probabilistically independent, we easily obtain

$$P_{H_1}(2W) = 1, \quad P_{H_2}(2W) = \frac{1}{4} \text{ and } P_{H_3}(2W) = 0$$

(H_3 is now effectively eliminated from consideration.)

Hence by Bayes' theorem (Theorem 4.3(b)), we have

$$P_{2W}(H_1) = \frac{P_{H_1}(2W)P(H_1)}{P_{H_1}(2W)P(H_1) + P_{H_2}(2W)P(H_2)} = \frac{4}{5}$$

and

$$P_{2W}(H_2) = \frac{P_{H_2}(2W)P(H_2)}{P_{H_1}(2W)P(H_1) + P_{H_2}(2W)P(H_2)} = \frac{1}{5}.$$

Clearly

$$P_{H_1 \cap 2W}(W_3) = 1 \text{ and } P_{H_2 \cap 2W}(W_3) = \frac{1}{2}.$$

Hence, by Theorem 4.5, we have

$$P_{2W}(W_3) = P_{H_1 \cap 2W}(W_3)P_{2W}(H_1) + P_{H_2 \cap 2W}(W_3)P_{2W}(H_2) = \frac{9}{10}.$$

[†] This is in fact an ill-posed problem as the sample space has not been given.

As it employs so many different techniques, this is a very interesting example to study. Note that the final answer depends on your willingness to assign equal probabilities to H_1, H_2 and H_3 at the outset.

Example 4.9 and the essential technique of its solution can be found on page 18 of *A Philosophical Essay on Probabilities*, which was first published in 1820 by the great French mathematician Pierre Simon Laplace.

Note: In Example 4.9 above, we have used a procedure that is very common in Bayesian analyses, namely the use of Bayes' theorem to upgrade the 'prior probabilities' $P(H_i)$ to the 'posterior probabilities' $P_{2W}(H_i)$. The probabilities $P_{H_i}(2W)$ are sometimes called 'likelihoods'. Theorem 4.3(a) is sometimes referred to as the 'principle of inverse probability' and is written in the form

$$\text{posterior probability} \propto (\text{likelihood} \times \text{prior probability}).$$

In general, we do not feel that an unrestricted use of the principle of sufficient reason is justified but we will continue to use the weaker principle of symmetry where appropriate. We note that symmetry arguments are particularly valuable for applications to science, where we are often, for example in statistical mechanics, dealing with models of reality comprising large numbers of 'particles' with identical properties. A modern and more powerful approach to the principle of symmetry which utilises the concept of entropy will be described in Section 6.4.

4.5.2 Subjective probabilities

Earlier we discussed subjective probabilities as being a measure of an individual's ignorance concerning a certain experience. The subjective theory of probability has been intensively developed in the twentieth century through the ideas of the British philosopher Frank Ramsey, the Italian mathematician Bruno de Finetti and the school of Bayesian statisticians. In the modern theory of subjective probability, a basic tenet is that *all probabilities are conditional* on the state of mind of the individual. For example, if we return to the game where Wales are playing Denmark in Section 4.2, there is no such thing as an 'absolute' probability that Wales will win the match. My personal probability that Wales will win is designated $P_H(A)$, where the hypothesis H contains all the relevant information that I am using to formulate my judgement about the game. For example, if I believe that Wales will win because their players will be better equipped to deal with the playing conditions and that the Danes will be seriously hampered by an injury to one of their best players, then these beliefs will all be included in H. If conditional probabilities are to be fundamental, the notion of a probability measure which we developed earlier will need to be amended. An appropriate notion of 'conditional probability space' is described by A. Rényi in his book *Probability Theory* (see references below) but we will not develop this notion here. Most of the results which we have obtained in this chapter can be extended to a form where they are suitable for using with subjective probabilities, for example the definition of probabilistic independence of A and B becomes

$$P_H(A \cap B) = P_H(A)P_H(B).$$

Unfortunately, the end result is that all the formulae we have developed for calculating with probabilities become cluttered up with extra terms, H, which play no further role in the analysis other than reminding us that P is subjective. Although I agree with many of the ideas of the subjectivists, my approach in this book will be to take H as understood and not display it manifestly unless it plays a direct role in calculations.

Finally, it is important not to misunderstand the sense of subjective probability and dismiss it as unscientific on the basis that it is 'all in the mind' or 'dependent on personal whim'. In fact, it is more appropriate to regard subjective probabilities as social rather than personal constructs. Two or more people may entertain different prior probabilities $P_{H_1}(A)$ and $P_{H_2}(A)$, respectively, about an experience A but, on updating their information, for example by engaging in rational discourse or carrying out new experiments (especially when engaged in scientific questions), they might eventually agree on the posterior probability of A. Many political and social questions can be analysed in terms of two or more organised groups disagreeing about the probability of an experience. For example, at various times since the invention of nuclear weapons, most recently in the early 1980s, we have seen societies in Western Europe unable to agree on the probability of imminent nuclear war.

4.5.3 The frequentist approach

We have already, in Section 4.2, examined and rejected the attempt to interpret probabilities in terms of relative frequencies. This approach to probability was systematically developed in the first half of the twentieth century by the philosopher Richard von Mises, and still has many adherents. Readers who are knowledgeable about modern Western philosophy might speculate that its popularity, at a time when logical positivism was dominant, lay with its prescription for verification of probabilities via experiment. If we reject the frequentist approach, we must still find a mechanism for assigning probabilities to events when the only information we have is the number of times they have occurred in a given number of identical experiments. The most convincing approach to this is via a Bayesian analysis.

Let us look at the simplest possible example. Suppose that we carry out successive tosses of a coin. Remember that we have no way of knowing at the outset whether or not the coin is fair but we may use the principle of insufficient reason to make the prior assignment $P(H) = \frac{1}{2}$. Suppose that after thirteen tosses we have observed eight heads (if the coin is indeed fair, it follows by use of the binomial distribution – see Section 5.7 below – that this event has a probability of 0.157). We may, however, suspect that the coin has a slight bias. What would be a reasonable assignment for the (subjective) probability that we obtain a head on the fourteenth throw given that the first thirteen throws have yielded eight heads? It seems that the only reasonable candidate for this probability is $\frac{8}{13}$; however, most probabilists would in fact regard this as somewhat naïve. A modern Bayesian approach would require us to make a prior assignment of the probability of the fourteenth throw being a head. The occurrence of eight heads out of the first thirteen is then regarded as *evidence* with which the prior probability can be upgraded to a posterior probability. There are precise mathematical procedures for calculating this, based on the important concept of exchangeability introduced by de Finetti – unfortunately the details are too complicated to be included here. However, it is interesting to observe that this theorem would lead us to assign the (highly non-intuitive) probability of $\frac{9}{15}$ in the above example – the same result would also be given by a formula obtained by Laplace in the nineteenth century and called the 'law of succession'.

An alternative approach to the relationship between probability and relative frequency is based on a result called the law of large numbers. We will examine this at length in Chapter 8.

We have looked above at three different approaches to probability. There are others which are too complex to discuss here, such as attempts to found probability as a system of inductive logic due, primarily, to J. M. Keynes (who is perhaps better known for his contributions to economics) and R. Carnap during this century.

I will close this section by summarising the approach to probability adopted in this book:

(i) Probability is a measure of total mass one, defined on the event in a sample space. As such it is a mathematical concept.
(ii) In using the concept of probability to analyse scientific models of aspects of reality, the correct choice of probability measure can often be found by symmetry considerations (or by entropy considerations – see Chapter 6).
(iii) When applying the concept of probability to complex real world experiences, all probabilities are conditional on the perspective of the individual or social group. Nonetheless, general agreement on probability assignments may often be reached as a result of rational discourse.

4.6 The historical roots of probability

Probability theory had a late start compared to other mathematical sciences, such as geometry and mechanics, which were highly developed by the ancient Greeks. Perhaps this is because the ancients saw the operation of chance in the universe as the activity of their gods and therefore not within the scope of human knowledge.

Another point is that when probability theory finally did begin to emerge, it was because of the needs of gamblers for more insight into the workings of their games. However, the Greeks and Romans gambled mainly by throwing four-sided objects called astralagi, which were made from bones and were not symmetrical. Consequently, ancient thinkers would not have had the stimulus to obtain the simplest rule for calculating probabilities, namely the principle of symmetry, by observing games of chance.[†]

The mathematical development of probability theory began in Renaissance Italy, and the man who perhaps deserves to be called the founder of the subject was the scholar and gambler Facio Cardano (1444–1524), whose manuscript *Liber de Ludo Aleae* (*Book on Games of Chance*) was found among his papers long after his death, and published in 1663. This contains the first formulation of the principle of symmetry within the context of dice games.

A number of important advances in combinatorial techniques were made in the seventeenth century. Perhaps the most important of these arose from the dialogue between the mathematician Pierre de Fermat (1601–1665) and the philosopher Blaise Pascal (1623–1662), which was initiated by problems arising from gambling posed by the Chevalier de Mere. The revolution in mathematical thought that developed out of the discovery of calculus by Newton and Leibniz had a significant impact on probability in the eighteenth century, and amongst those who applied the new techniques to probability were Pierre de Montfort (1678–1719) and Abraham de Moivre (1667–1754). Thomas Bayes (1702–1761) published his famous theorem in 1763.

In the nineteenth century, calculus matured into mathematical analysis and this, too, began to be applied within probability theory. A great landmark was the *Théorie analytique des Probabilités* by Pierre Simon, Marquis de Laplace (1749–1827). The introduction to this work was published separately as the *Essai Philosophique sur les Probabilités* in 1820 and gives a coherent and systematic presentation of the subject, as it stood at the time, for the general reader.

As the nineteenth century progressed, probability theory received a myriad of stimuli from the emerging study of statistics as well as from mathematics itself. The twentieth century will perhaps be remembered as the age of abstraction in mathematics, with its emphasis on mathematical structures and the relationships between them. An important step in the development of probability theory was the publication by the great Russian mathematician A. N. Kolmogorov (1903–1987) of his *Foundations of the Theory of Probability* in 1933, in which the axioms for a probability measure were first introduced.

[†] It may be that this brief account is overly 'eurocentric'. There are indications that crude probabilistic ideas were used in ancient times in India – for more about this (and references) see pages 3–4 of M. M. Rao and R. J. Swift *Probability Theory with Applications* (Springer, 2006).

Exercises

4.1. How would you assign probabilities to the following experiences? You might use the principle of symmetry, subjective probability, relative frequency or a combination of these:

   (a) obtaining a flush in a game of poker,
   (b) team A beating team B in a football game, given that A have won nine of the last 17 matches between these teams,
   (c) the sun will rise in the East tomorrow morning,
   (d) the claim that smoking causes cancer.

4.2. Assign personal probabilities to each of the following propositions:

   (a) The party presently in government will win the next General Election.
   (b) You will set up home with a new partner within the next year.
   (c) The world will end in July 2999.

4.3. Three fair coins are tossed:

   (a) Write down the sample space.
   (b) What are the events:

      (i) the first coin is a head (E),
      (ii) the second coin is a tail (F).

   (c) Calculate the probabilities $P(E \cup F)$, $P(E \cap F)$, $P(\bar{E})$ and $P(\bar{F})$.

4.4. If two fair die are rolled, what is the probability that the sum of the upturned faces will be seven?

4.5. Two events A and B are such that $P(A) = 0.45$, $P(B) = 0.22$ and $P(A \cup B) = 0.53$. Find $P(\bar{A} \cup \bar{B})$.

4.6. A company manufacturing silicon chips classifies them as follows for quality control:

   A – in perfect condition,
   B – containing minor defects which will not inhibit operation,
   C – major defects.

   In a batch of 170 items, 102 are of type A and 43 of type B. If a chip is chosen at random, find the probability that:

   (i) it is of type C,
   (ii) it will be operational.

4.7. If you buy a copy of the new album 'Fell Over My Brain' by Sonic Delirium, you get a free picture of the band with probability 0.15 and an extra track with probability 0.21. Assuming that these are independent, find the probability of obtaining:

   (a) a free picture with no extra track,
   (b) an extra track but no free picture,
   (c) both a free picture and an extra track,
   (d) at least one of these.

4.8. Convince yourself that the Dirac measure (as described in Example 3.8) is a probability measure.

4.9. Let $S = \{x_1, x_2, \ldots, x_n\}$ be a finite set and f a real valued function on S for which each $f(x_i) \geq 0$. Show that if

$$P(x_i) = \frac{f(x_i)}{Z}$$

where $Z = \sum_{i=1}^{n} f(x_i)$, then P is a probability measure on $\mathcal{B}(S)$. [*Hint*: Use the result of Exercises 3.14.]

    Note: An important application of this probability measure is to statistical mechanics. Here the x_is are the possible energies available to be taken up by a large collection of particles. In this case we have

$$f(x_i) = e^{-\beta x_i}$$

where β is a positive constant; Z is called the partition function and P is called the Gibbs distribution (see Sections 6.4–6.5 to learn more about this).

4.10. Let $A \in \mathcal{P}(S)$ with $P(A) \neq 0$. Show that the conditional probability P_A defines a probability measure on $\mathcal{P}(A)$.

4.11. Show that $P_A(B) = P_A(A \cap B)$.

4.12. If A and B are disjoint with $P(A) \neq 0$ and $P(B) \neq 0$, show that

$$P_{A \cup B}(C) = \beta P_A(C) + (1 - \beta) P_B(C)$$

where $\beta = \frac{P(A)}{P(A) + P(B)}$. (*Note*: This example shows that for fixed C, if we make the prescription $\rho(A) = P_A(C)$, where $A \in \mathcal{B}(S)$, then ρ is NOT a measure.)

4.13. If two balls are randomly drawn (without replacement) from a bowl containing seven red and five blue balls, what is the probability that:

  (i) both balls are red?
  (ii) the first ball is red and the second blue?
  (iii) one ball is red and the other blue?

4.14. In a factory producing compact discs, the total quantity of defective items found in a given week is 14%. It is suspected that the majority of these come from two machines, X and Y. An inspection shows that 8% of the output from X and 4% of the output from Y is defective. Furthermore, 11% of the overall output came from X and 23% from Y. A CD is chosen at random and found to be defective. What is the probability that it came from either X or Y?

4.15. Consider again the binary symmetric channel discussed in the text. What values do you obtain for $Q(0)$, $Q(1)$ and the conditional probabilities $Q_i(j)(i, j, = 1, 2)$ when $p = \frac{1}{2}$ so that there is maximum uncertainty at the input? Interpret your results.

4.16. In a binary erasure channel, some characters are erased before reaching the output. The set-up is as shown in Fig. 4.4. where E indicates erasure. Calculate:

  (a) the output probabilities $Q(0)$, $Q(E)$ and $Q(1)$,
  (b) the conditional probabilities $Q_0(0)$, $Q_E(0)$, $Q_E(1)$, $Q_1(1)$.

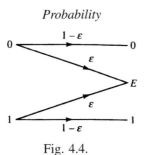

Fig. 4.4.

4.17. A combination of the binary symmetric and binary erasure channels is that given in Fig. 4.5. Calculate all the output probabilities and conditional probabilities $Q_i(j)$.

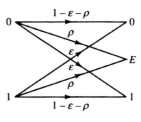

Fig. 4.5.

4.18. Three messages are sent across a channel. The probabilities of successful reception are 0.57, 0.69 and 0.93 respectively. Assuming that the signals are all probabilistically independent, find the probability of successfully receiving:

(a) all three signals,
(b) no signals,
(c) two signals only.

4.19. Two events may be probabilistically independent but not disjoint. To see this let $S = \{1, 2, 3, 4\}$, with each singleton set having probability 0.25. Let $A = \{1, 4\}$ and $B = \{2, 4\}$. Show that $P(A \cap B) = P(A)P(B)$.

4.20. Suppose that two fair dice are thrown. Consider the events A, B and C defined as follows:

A: 1, 2 or 3 is thrown on the first die.
B: 4, 5 or 6 is thrown on the second die.
C: The numbers on both dice are either less than or equal to 3 or they are greater than or equal to 4.

Show that each pair of events is independent but the three events A, B and C are not independent, that is

$$P(A \cap B \cap C) \neq P(A)P(B)P(C).$$

4.21. Show that if any three events A, B and C are independent, then $A \cup B$ is independent of C.

4.22. Two binary symmetric channels are connected in series, as is shown in Fig. 4.6, so that the output at B from the first channel is transmitted to C along the second channel.

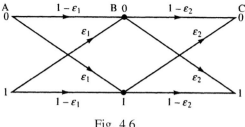

Fig. 4.6.

Assuming that the two channels are probabilitistically independent, calculate the output probabilities at C, given that the input probabilities are determined by $p = 0.4$ (where p is the probability that a 1 is sent out) with $\varepsilon_1 = 0.14$ and $\varepsilon_2 = 0.21$.

4.23. A set of odds is said to be coherent if the associated probabilities add up to 1 (which they must do if they really are probabilities). Three friends A, B and C are to run a race to see who is fastest. A fourth friend offers odds of 3 to 1 against A, 7 to 1 against B and evens on C. Are these odds coherent? If not, how would you alter the odds on C to make them so.

4.24. The famous mathematician Jean d'Alembert (1717–1783) argued that for a fair coin tossed twice in succession, the probability of a head was $\frac{2}{3}$. He claimed that the correct sample space was {H, TH, TT} as, if a head appeared on the first toss, there would be no need to toss it a second time to obtain the head. Criticise this argument.

4.25. The following problem first appeared in the Fermat–Pascal correspondence. What number (n) of throws of a fair die has to be made before the probability of at least one 6 is $\frac{1}{2}$?

4.26. Fermat posed the following problem, which was first solved by Christianus Huygens (1629–1695). Given forty cards of which ten are red, ten blue, ten green and ten purple, what is the probability of picking four cards at random (without replacement) which are each of different colours?

4.27. Find the probability that five cards chosen at random from a deck of 52 will contain:

  (a) no aces,
  (b) at least one ace,
  (c) exactly one ace.

4.28. In the British National Lottery, you attempt to win prizes by choosing six different numbers between 1 and 49 (inclusive) and matching these to the numbers showing on six coloured balls emitted at random from 49. To win the jackpot you have to get all six numbers correct. To win the second prize, you have to have five of your six numbers correct and also match your sixth number to that on an additional seventh (bonus) ball emitted. There are also prizes for getting five, four and three numbers correct. Find the probability of: (a) winning the jackpot, (b) getting five numbers only correct, (c) winning the second prize, (d) getting four numbers only correct, (e) getting three numbers only correct, (f) winning a prize at all, (g) getting no numbers correct.

4.29. Suppose that you go to a doctor because you believe that you are suffering from some disease. The doctor gives you a test and you register positive. Moreover, the test is known to be highly sensitive in that

$$P_{D_p}(T_p) = P_{D_a}(T_n) = 0.99$$

where $D_p(D_a)$ is the event that the disease is present (absent) and $T_p(T_n)$ is the event that you test positive (negative). You may now conclude that you are indeed suffering from the disease in question; however, this may not be justified for you have failed to take into account the incidence of the disease in the population. Convince yourself that the correct indicator as to your likelihood of having the disease is $P_{T_p}(D_p)$ and calculate this when $P(D_p) = 0.0001$. (This example was taken from the article 'Common-sense and statistics' by Adrian Smith in *Teaching and Using Statistics*, edited by Neville Davies (Royal Statistical Society, 1994).)

Further reading

There are an enormous number of introductory textbooks available on probability theory. The best of these, beyond doubt, is William Feller's *An Introduction to Probability Theory and its Applications, Volume 1* (2nd edition, J. Wiley and Sons, 1964). A less well-known but beautifully written and comprehensive text is A. Rényi, *Probability Theory* (North-Holland, 1970). Also thoroughly recommended is G. Grimmett and D. Welsh, *Probability – An Introduction* (Clarendon Press, Oxford, 1986). One of the most important and influential books in the development of modern probability is A. N. Kolmogorov's *Foundations of the Theory of Probability* (English translation, Chelsea, New York, 1956), which is very short, succinct and highly readable. Kolmogorov was one of the greatest mathematicians of the twentieth century and an innovator in many fields. Some insight into his achievements can be obtained from his obituary by K. R. Parthasarathy in the *Journal of Applied Probability* **25**, 445–50 (1988). A highly fascinating and more down to earth treatment of probability than the above can be found in *The Art of Probability for Engineers and Scientists* by R. W. Hamming (Addison Wesley, 1991). Hamming is one of the major contributors to the modern theory of coding.

For a subjective approach to probability theory, the best introduction is D. V. Lindley *Introduction to Probability and Statistics from a Bayesian Viewpoint, Part 1. Probability* (Cambridge University Press, 1969). De Finetti's highly original *Probability Theory*, in two volumes, has been reissued by J. Wiley and Sons (1990).

A nice overview of the different interpretations of probability can be found in R. Weatherford's *Philosophical Foundations of Probability Theory* (Routledge and Kegan Paul, 1982). For those who hanker after a frequentist interpretation, D. A. Gillies' *An Objective Theory of Probability* (Methuen, 1973) is very readable. *Creating Modern Probability* by J. von Plato (Cambridge University Press, 1994) gives a fascinating historical survey of the various approaches to the foundations of probability which have been developed during the twentieth century.

Much that I have learned about the history of probability can be found in the excellent *Games, Gods and Gambling* by F. N. David (Charles Griffin and Co., 1962). Bayes original paper and much else of interest besides is in *Studies in the History of Statistics and Probability* edited by E. S. Pearson and M. G. Kendall (Charles Griffin and Co., 1970). Laplace's *A Philosophical Essay on Probabilities* has been reissued by Dover Publications (1951) and is quite accessible to modern readers.

Finally, a highly entertaining and thought-provoking novel based on the use of chance to explore 'human potential' is *The Dice Man* by Luke Rhinehard (Granada Publishing, Panther, 1972).

5

Discrete random variables

5.1 The concept of a random variable

We have, in the last chapter, become used to studying problems involving chance using the probability space $(S, \mathcal{B}(S), P)$, where S is a set, the sample space, whose elements are the possible outcomes of some experience; $\mathcal{B}(S)$ is a Boolean algebra whose elements, the events, represent all the propositions about the outcomes which it is possible to place bets on; and P is a probability measure on S. Now suppose that $\#(S) = n$ and take $\mathcal{B}(S) = \mathcal{P}(S)$ so that by Theorem 3.1, $\#\mathcal{P}(S) = 2^n$. Clearly, as n becomes large, 2^n is very large and soon becomes quite an unwieldy number (e.g. when $n = 10$, $2^n = 1024$). Furthermore, in many practical problems, it turns out that we only need to examine a small number of these events. Clearly, it would be useful to have a device which would allow us to model our experience without having to exhibit all the events in $\mathcal{B}(S)$. Such a device is provided by the theory of random variables.

A *discrete random variable X* is just a mapping, from S into a subset R of $\mathbb{R}$ where $\#(R) = n$ or ∞, that is it is a function which attaches a number x to each outcome in S. The set R is called the *range* of X. The term 'discrete' is included because of the restriction on R (in Chapter 8 we will study continuous random variables, where this restriction is removed). In many of the examples we will study below, R will be a subset of the natural numbers $\mathbb{N}$.

Note: In a more general framework, we should restrict the notion of random variables to those mappings satisfying an additional property called *measurability*. If, as in many of the examples given below, we have $\mathcal{B}(S) = \mathcal{P}(S)$, this is automatically satisfied. We will discuss this again in Chapter 8.

Effectively, we may think of X as a variable whose value is uncertain and, in practice, the main interest will be in establishing the probability that a certain value of X arises. The example given below should clarify these ideas.

Example 5.1 Two fair dice are thrown. What is the probability that the sum of the numbers appearing on the two upturned faces equals 7?

Solution We write each outcome as a pair of numbers

(score on die 1, score on die 2).

As each die has six faces, the basic principle of counting tells us that S has 36 members so that $\#\mathcal{P}(S) = 2^{36} = 6.871\,947\,7 \times 10^{10}$! By the principle of symmetry, we see that each outcome has probability $\frac{1}{36}$. We now introduce a random variable X:

$$X = \text{sum of scores of the two dice.}$$

Clearly, X can take any whole number between 2 and 12 as its value. We are interested in the event that X takes the value 7. We write this simply as $(X = 7)$, then

$$(X = 7) = \{(1, 6), (2, 5), (3, 4), (4, 3), (5, 2), (6, 1)\}.$$

Now by (P8), we have $P(X = 7) = \frac{6}{36} = \frac{1}{6}$.

If we were betting on the game of throwing two dice, we might want to establish which of the numbers from 2 to 12 has the highest probability of occurring. This gives us the 'safest bet'.

Example 5.2 Two fair dice are thrown. Which sum of scores on the two dice has the maximum probability?

Solution To establish this, we calculate all the probabilities $P(X = r)$ where r goes from 2 to 12, imitating the process of Example 5.1. We find (check) that

$$P(X = 2) = \frac{1}{36} \qquad\qquad P(X = 12) = \frac{1}{36}$$

$$P(X = 3) = \frac{1}{18} \qquad\qquad P(X = 11) = \frac{1}{18}$$

$$P(X = 4) = \frac{1}{12} \qquad\qquad P(X = 10) = \frac{1}{12}$$

$$P(X = 5) = \frac{1}{9} \qquad\qquad P(X = 9) = \frac{1}{9}$$

$$P(X = 6) = \frac{5}{36} \qquad\qquad P(X = 8) = \frac{5}{36}$$

$$P(X = 7) = \frac{1}{6}$$

so we see that $X = 7$ is the best bet.

We observe that the use of the random variable X has allowed us to sift effortlessly through the 2^{36} events in $\mathcal{P}(S)$ and isolate the 11 which are relevant to this problem.

As we have seen in Example 5.2 above, the list of numbers $P(X = r)$ plays a major role in the theory of random variables. We study these in greater detail in the next section.

5.2 Properties of random variables

Let X be a random variable on S taking values in the set $R = \{x_1, x_2, \ldots, x_n\} \subset \mathbb{R}$. Now define the numbers

$$p(x_j) = P(X = x_j) \quad \text{for } 1 \leq j \leq n.$$

p is called the *probability distribution* or *probability law* of X. For example, in Example 5.2 above, $R = \{2, 3, \ldots, 12\}$ and $p(2) = \frac{1}{36}$, $p(3) = \frac{1}{18}$, etc.

Now consider the Boolean algebra $\mathcal{P}(R)$ and for $A \in \mathcal{P}(R)$ define

$$p(A) = \sum_{x_j \in A} p(x_j)$$

where the sum is over those x_js which lie in A.

Lemma 5.1 *p is a probability measure on $\mathcal{P}(R)$.*

Proof Since each $p(x_j) \geq 0$, it follows from Exercise 3.14 that p is a measure. To show that it is a probability measure, we have

$$p(R) = \sum_{j=1}^{n} p(x_j) = \sum_{j=1}^{n} P(X = x_j) = 1 \qquad \text{by (P8)}$$

since the events $(X = x_1), (X = x_2), \ldots, (X = x_n)$ form a partition of S. $\qquad \square$

Note

(i) For those who have studied mappings on sets we can characterise p as the probability measure on $\mathcal{P}(R)$ given by

$$p = P \circ X^{-1}.$$

(ii) Although, in the above example, the sum which defines $p(A)$ is finite, if R has an infinite number of elements, then the corresponding sum will be infinite and should thus be interpreted as a convergent series.

In the proof of Lemma 5.1, we established the following useful formula

$$\sum_{j=1}^{n} p(x_j) = 1. \tag{5.1}$$

A graph of p, called its *probability histogram*, gives a quick visual insight into the behaviour of X. Figure 5.1 is the histogram for the distribution found in Example 5.2 above.

In the sequel, we will often encounter events such as $(X \leq x_r)$, where $1 \leq r \leq n$. This is precisely the set of all those outcomes for which the value of X is less than or equal to x_r, for example in Example 5.1 above

$$(X \leq 4) = \{(1, 1), (1, 2), (1, 3), (3, 1), (2, 1), (2, 2)\}.$$

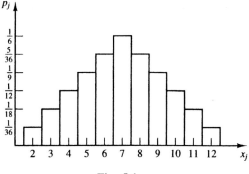

Fig. 5.1.

The events $(X < x_r)$, $(X \geq x_r)$, $(X > x_r)$ $(x_r \leq X \leq x_s)$, etc. are defined similarly.

A very useful function which we can associate to a random variable X is its *cumulative distribution* F. This is defined by

$$F(x_r) = P(X \leq x_r)$$

for $1 \leq r \leq n$, from which the reader should easily deduce that

$$F(x_r) = \sum_{j=1}^{r} p(x_j). \tag{5.2}$$

The following lemma collects together some useful properties of F.

Lemma 5.2

(i) $F(x_{r+1}) - F(x_r) = p(x_{r+1})$ *for* $1 \leq r < n$.

(ii) F *is an increasing function, that is*

$$F(x_{r+1}) \geq F(x_r) \quad \text{for } 1 \leq r < n.$$

(iii) $F(x_n) = 1$.

Proof

(i) As $(X \leq x_{r+1}) = (X \leq x_r) \cup (X = x_{r+1})$, we have

$$P(X \leq x_{r+1}) = P(X \leq x_r) + P(X = x_{r+1})$$

by (P3) and the result follows.

(ii) follows immediately from (i) as each $p(x_{r+1}) \geq 0$.

(iii) follows from (5.2) and (5.1).

□

The cumulative distribution is sometimes more useful than the probability law p in solving practical problems, as we will see below. For that reason, it is often lists of values of F which appear in 'statistical tables'.

Example 5.3 Find the cumulative distribution for the random variable of Example 5.2.

Solution

$$F(2) = \frac{1}{36}, \ F(3) = \frac{1}{12}, \ F(4) = \frac{1}{6}, \ F(5) = \frac{5}{18}, \ F(6) = \frac{5}{12},$$

$$F(7) = \frac{7}{12}, \ F(8) = \frac{13}{18}, \ F(9) = \frac{5}{6}, \ F(10) = \frac{11}{12},$$

$$F(11) = \frac{35}{36}, \ F(12) = 1.$$

The graph of F is shown in Fig. 5.2.

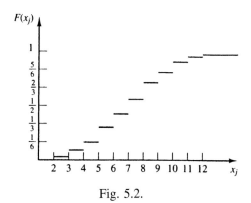

Fig. 5.2.

The following results are sometimes useful in solving specific problems, as we will see below.

Lemma 5.3

(i) *For $1 \le r \le n$, we have*

$$P(X > x_r) = 1 - F(x_r).$$

(ii) *For $1 \le r < s \le n$, we have*

$$P(x_r \le X \le x_s) = F(x_s) - F(x_{r-1}).$$

Proof

(i) Since $(X > x_r) = \overline{(X \le x_r)}$, the result follows from (P5).
(ii) As $(x_r \le X \le x_s) = (X \le x_s) - (X \le x_{r-1})$, the result follows from (P4).

☐

To illustrate the value of Lemma 5.3, consider the following.

Example 5.4 Return again to the set-up of Examples 5.1 and 5.2. What is the probability that the combined score of the two die is: (a) not less than 8, (b) between 5 and 9 (inclusive).

Solution

(a) We require $P(X \geq 8) = P(X > 7)$

$$= 1 - F(7) \qquad \text{by Lemma 5.3(i)}$$

$$= 1 - \frac{7}{12} \qquad \text{by Example 5.3}$$

$$= \frac{5}{12}.$$

(b)

$$P(5 \leq X \leq 9) = F(9) - F(4) = \frac{5}{6} - \frac{1}{6} = \frac{2}{3}$$

by Lemma 5.3(ii) and Example 5.3.

There are some random variables which appear so often in different contexts that they have their own names. We briefly consider three of the simplest of these and will return in Sections 5.7–5.8 to study some more complex examples.

5.2.1 The discrete uniform random variable

This is a random variable X for which

$$p(x_1) = p(x_2) = \cdots = p(x_n) = \frac{1}{n}.$$

Uniform random variables arise naturally when we apply the principle of symmetry, and we have already done a lot of work on these in the preceding chapter, although we did not use the terminology of random variables there. An example is throwing a single fair die and letting

$$X = \text{number on uppermost face of die.}$$

In this case $R = \{1, 2, 3, 4, 5, 6\}$ and $n = 6$.

5.2.2 Bernoulli random variables

These are named after Jacob Bernoulli, who in the eighteenth century wrote an important manuscript about probability theory. We take $R = \{0, 1\}$ and define

$$p(1) = p, \; p(0) = 1 - p \qquad \text{where } 0 \leq p \leq 1.$$

Examples of such random variables include the tossing of a biased coin, choosing an experience at random, which might be a 'success' or 'failure', or emitting symbols into a binary symmetric channel (see Section 4.3).

A Bernoulli random variable is uniform if and only if $p = \frac{1}{2}$ so that both 0 and 1 are equiprobable. This special case is often called the *symmetric Bernoulli distribution*.

5.2.3 'Certain' variables

Suppose that out of a range of possibilities $\{x_1, x_2, \ldots, x_n\}$, we know for certain that a particular value, x_j say, will occur. It is often useful to consider x_j as the value of a random variable X whose probability law is the Dirac measure δ_{x_j} at x_j (recall Example 3.8).

Some readers may wonder why we haven't specified the probability space in the above examples. This is because it has, to some extent, become irrelevant and all the information we require is contained in the random variable and its distribution.

Example 5.5 A box contains five red balls and seven blue ones. In all other respects the balls are identical. What is the probability that a randomly chosen ball is red? What is an appropriate random variable to describe this situation?

Solution We have a sample space S containing 12 members. The event R that we choose a red ball has five members, so, by the principle of symmetry, the required probability is $p = \frac{5}{12}$. We now introduce the random variable X with the interpretation $1 \rightarrow$ 'red' and $0 \rightarrow$ 'blue'; then X is Bernoulli with $p = \frac{5}{12}$.

Next we consider the 'algebra of random variables'. Random variables can be added, multiplied by 'scalars' and multiplied together to form new random variables. More precisely, let X and Y be two random variables and α be a real number; then we can form new random variables $X + Y$, αX, $X + \alpha$ and XY. To understand the meaning of these operations, we consider some examples. For example:

 (i) Let X be the number of boys born in your city next year and Y be the number of girls; then $X + Y$ is the number of children born next year.
 (ii) Let X denote the length (in feet) of a randomly chosen wooden plank removed from a lorryload. Let $\alpha = 0.3038$; then αX denotes the same length in metres.
(iii) If X is as in (ii) and α is the known length of a steel cap to be placed on the edge of each plank, then $X + \alpha$ is the length of a (randomly chosen) capped plank.
(iv) Let X be a random force acting on a particle and Y be a random distance that the particle might move; then XY is the (random) work done on the particle.

Clearly, we can multiply random variables by themselves to form X^2, X^3, etc. This suggests that we might be able to form the random variable $f(X)$, where f

is any function on $\mathbb{R}$. In fact this is so; if X has range $\{x_1, x_2, \ldots, x_n\}$, then $f(X)$ has range $\{f(x_1), f(x_2), \ldots, f(x_n)\}$. The following, rather contrived, example illustrates this idea. For example, let X be a random variable taking the values $\{0, \pi/4, \pi/2\}$ with probabilities $\frac{1}{4}, \frac{1}{2}$ and $\frac{1}{4}$, respectively; then $\sin(X)$ takes the values $\{0, 1/\sqrt{2}, 1\}$ with probabilities $\frac{1}{4}, \frac{1}{2}$ and $\frac{1}{4}$ respectively.

Note: Some readers might appreciate a formal definition of the above operations. We recall that X and Y are functions on S. For each $s \in S$, we then define

$$(X + Y)(s) = X(s) + Y(s),$$

$$(\alpha X)(s) = \alpha X(s),$$

$$(XY)(s) = X(s)Y(s),$$

$$f(X)(s) = f(X(s)).$$

Clearly, the probability laws of αX and $f(X)$ (if f is one-to-one) are the same as that of X, but the determination of the laws of $X + Y$ and XY is more problematic and, in fact, a general formula for the former of these can only be established in a special case (see below, Section 5.5).

We conclude this section by establishing a useful result about the 'joint distribution' of two random variables. More specifically, let X be a random variable with range $R_X = \{x_1, x_2, \ldots, x_n\}$ and probability law p and let Y be a random variable (with the same sample space as X) with range $R_Y = \{y_1, y_2, \ldots, y_m\}$ and probability law q. We define the *joint distribution* of X and Y by

$$p_{ij} = P((X = x_i) \cap (Y = y_j)) \tag{5.3}$$

for $1 \le i \le n, 1 \le j \le m$.

Lemma 5.4

(i) $\displaystyle\sum_{i=1}^{n} p_{ij} = q_j \qquad$ *for each* $1 \le j \le m$.

(ii) $\displaystyle\sum_{j=1}^{m} p_{ij} = p_i \qquad$ *for each* $1 \le i \le n$.

(iii) $\displaystyle\sum_{i=1}^{n} \sum_{j=1}^{m} p_{ij} = 1$.

Proof (i) Apply Exercise 3.16 taking $A = (Y = y_j)$ and $(X = x_1), (X = x_2), \ldots,$ $(X = x_n)$ as the partition of S. Item (ii) is proved similarly. Item (iii) follows on applying (5.1) to (i) or (ii). $\qquad\square$

Note: $\{p_{ij}, 1 \le i \le n, 1 \le j \le m\}$ is the probability law of the random vector $(X, Y) = X\mathbf{i} + Y\mathbf{j}$ (see Chapter 9).

5.3 Expectation and variance

Recall Example 4.3 concerning the probability of Arctic Circle winning the 3.15 at Haydock Park. We calculated this probability to be $\frac{2}{9}$. Now suppose that I bet £5 on Arctic Circle and the horse wins the race. How much do I expect to win? Clearly, this is £5 $\times \frac{2}{9}$ = £1.11 p. Such considerations led the founders of probability to define the expectation of an event A, $\mathbb{E}(A)$ by the formula

$$\mathbb{E}(A) = (\text{amount staked}) \times P(A).$$

In the modern theory, the key notion is the expectation $\mathbb{E}(X)$ of a random variable X. To justify the definition of $\mathbb{E}(X)$ given below, imagine that you place bets on the outcomes of X as follows; you bet x_1 on the event $(X = x_1)$, x_2 on the event $(X = x_2), \ldots, x_n$ on the event $(X = x_n)$, then your expected winnings would be

$$\mathbb{E}(X) = \sum_{j=1}^{n} x_j p_j. \tag{5.4}$$

For example, suppose that $R = \{1, 2, 3\}$ with $p(1) = \frac{1}{6}$, $p(2) = \frac{1}{3}$ and $p(3) = \frac{1}{2}$; then

$$\mathbb{E}(X) = \left(1 \times \frac{1}{6}\right) + \left(2 \times \frac{1}{3}\right) + \left(3 \times \frac{1}{2}\right) = \frac{7}{3}.$$

Some insight into the meaning of $\mathbb{E}(X)$ may be gained by considering a relative frequencies approach. Suppose in the above example we repeat the experiment represented by X twelve times; then we 'expect' to obtain the number 1 twice, the number 2 four times and the number 3 six times, in which case we see that $\mathbb{E}(X)$ is precisely the average (or mean) value of the expected results, that is

$$\mathbb{E}(X) = \frac{(2 \times 1) + (4 \times 2) + (6 \times 3)}{12} = \frac{7}{3}.$$

As a result of these considerations, $\mathbb{E}(X)$ is sometimes called the *mean* of the random variable X. We should not, however, be deluded into believing that if we carry out the above experiment 12 times, we will in fact get a mean of $\frac{7}{3}$. More insight into how the values of the mean 'fluctuate' around $\mathbb{E}(X)$ will be gained in Chapter 8.

Note: If the range of X is infinite, we shall assume in this chapter that X is such that the infinite series (5.4) converges.

Example 5.6 Find the mean of the random variable X whose values are the sums of the scores on two fair dice.

Solution Using the data of Example 5.2, we find that

$$\mathbb{E}(X) = \left(2 \times \frac{1}{36}\right) + \left(3 \times \frac{1}{18}\right) + \left(4 \times \frac{1}{12}\right) + \left(5 \times \frac{1}{9}\right)$$

$$+ \left(6 \times \frac{5}{36}\right) + \left(7 \times \frac{1}{6}\right) + \left(8 \times \frac{5}{36}\right) + \left(9 \times \frac{1}{9}\right)$$

$$+ \left(10 \times \frac{1}{12}\right) + \left(11 \times \frac{1}{18}\right) + \left(12 \times \frac{1}{36}\right) = 7.$$

Exercise: Throw two dice and calculate your average score after 10 throws, 20 throws, ..., 100 throws. As you increase the number of throws is the average getting 'closer' to 7?

Example 5.7 Find $\mathbb{E}(X)$ when:

(a) X is Bernoulli,
(b) X is uniform,
(c) $X = x_k$ with certainty.

Solution

(a) $\mathbb{E}(X) = (1 \times p) + (0 \times (1 - p)) = p$,
(b) $\mathbb{E}(X) = \sum_{j=1}^{n} x_j \times \frac{1}{n} = \frac{1}{n} \sum_{j=1}^{n} x_j$,
(c) $\mathbb{E}(X) = \sum_{j=1}^{n} x_j \delta_{x_k} = x_k$.

Note that in (b) we obtain the well-known arithmetic mean of the numbers $\{x_1, x_2, \ldots, x_n\}$.

If X and Y are two random variables with joint distribution (5.3), you should convince yourself that it makes sense to define

$$\mathbb{E}(X + Y) = \sum_{i=1}^{n} \sum_{j=1}^{m} (x_i + y_j) p_{ij}$$

and

$$\mathbb{E}(XY) = \sum_{i=1}^{n} \sum_{j=1}^{m} (x_i y_j) p_{ij}.$$

We now collect some useful facts about $\mathbb{E}(X)$. We begin with a definition. A random variable X is said to be *non-negative* if $R \subset [0, \infty)$, that is no value in its range is negative. Note that X^2 is always non-negative irrespective of whether X is.

Theorem 5.5

(a) *Let X and Y be two random variables defined on the same sample space, then*

$$\mathbb{E}(X + Y) = \mathbb{E}(X) + \mathbb{E}(Y).$$

(b) *Let X be a random variable and $\alpha \in \mathbb{R}$, then*

$$\mathbb{E}(\alpha X) = \alpha \mathbb{E}(X)$$

and

$$\mathbb{E}(X + \alpha) = \mathbb{E}(X) + \alpha.$$

(c) *Let X be a non-negative random variable, then*

$$\mathbb{E}(X) \geq 0.$$

Proof

(a) We use the notation of Lemma 5.4; then

$$\mathbb{E}(X + Y) = \sum_{i=1}^{n} \sum_{j=1}^{m} (x_i + y_j) p_{ij}$$

$$= \sum_{i=1}^{n} x_i \sum_{j=1}^{m} p_{ij} + \sum_{j=1}^{m} y_j \sum_{i=1}^{n} p_{ij}$$

$$= \sum_{i=1}^{n} x_i p_i + \sum_{j=1}^{m} y_j q_j$$

$$= \mathbb{E}(X) + \mathbb{E}(Y) \qquad \text{by Lemma 5.4(i) and (ii).}$$

(b)

$$\mathbb{E}(\alpha X) = \sum_{i=1}^{n} \alpha x_i p_i = \alpha \sum_{i=1}^{n} x_i p_i = \alpha \mathbb{E}(X),$$

$$\mathbb{E}(X + \alpha) = \sum_{i=1}^{n} (x_i + \alpha) p_i = \sum_{i=1}^{n} x_i p_i + \alpha \sum_{i=1}^{n} p_i$$

$$= \mathbb{E}(X) + \alpha \qquad \text{by (5.1).}$$

(c) Immediate from (5.4) since each $x_j \geq 0$, $p_j \geq 0$.

$\square$

Note: From (a) and (b) it follows that $\mathbb{E}$ is a linear functional on the space of random variables, that is for all α, $\beta \in \mathbb{R}$

$$\mathbb{E}(\alpha X + \beta Y) = \alpha \mathbb{E}(X) + \beta \mathbb{E}(Y).$$

In the following, we will occasionally adopt the standard notation of denoting $\mathbb{E}(X)$ as μ. Now, if f is a real-valued function, we have

$$\mathbb{E}(f(X)) = \sum_{j=1}^{n} f(x_j) p_j.$$

In particular, take $f(X) = (X - \mu)^2$; then $f(X)$ is non-negative, so by Theorem 5.5(c), $\mathbb{E}((X - \mu)^2) \geq 0$. We define the *variance* of X, $\mathrm{Var}(X)$ by

$$\mathrm{Var}(X) = \mathbb{E}((X - \mu))^2$$

$$= \sum_{j=1}^{n} (x_j - \mu)^2 p_j. \tag{5.5}$$

The interpretation of $\mathrm{Var}(X)$ is that it gives a measure of the extent to which we expect the result of the experiment represented by X to deviate, on average, from the mean μ. Clearly, the deviation of each x_j from μ is $(x_j - \mu)$. We might argue that an obvious candidate to measure 'average deviation' is

$$\sum_{j=1}^{n} (x_j - \mu) p_j = \mathbb{E}(X - \mu),$$

but this is clearly 0 by Theorem 5.5(b). Alternatively, we might consider $\mathbb{E}(|X - \mu|)$ but we prefer to use (5.5) as (see below) it has much better mathematical properties. The only problem with (5.5) is the practical one that if the values of X are in (units), then $\mathrm{Var}(X)$ is measured in (units)^2. For this reason, we find it useful to introduce the *standard deviation* $\sigma(X)$ of the random variable X by

$$\sigma(X) = \sqrt{(\mathrm{Var}(X))}. \tag{5.6}$$

We will sometimes simply denote $\sigma(X)$ as σ and $\mathrm{Var}(X)$ as σ^2. The following result gives some useful properties of the variance.

Theorem 5.6

(a) $\mathrm{Var}(X) = \mathbb{E}(X^2) - (\mathbb{E}(X))^2$.
(b) $\mathrm{Var}(\alpha X) = \alpha^2 \mathrm{Var}(X)$, *for all $\alpha \in \mathbb{R}$.*

Proof

(a) Expand $(X - \mu)^2 = X^2 - 2\mu X + \mu^2$, so by Theorem 5.5(a) and (b) (see also Exercise 5.6) we have

$$\mathrm{Var}(X) = \mathbb{E}(X)^2 - 2\mu\mathbb{E}(X) + \mu^2$$

from which the result follows.

(b)

$$\text{Var}(\alpha X) = \mathbb{E}((\alpha X - \mathbb{E}(\alpha X))^2)$$
$$= \mathbb{E}((\alpha X - \alpha\mu)^2) \qquad\qquad \text{by Theorem 5.5(b)}$$
$$= \mathbb{E}(\alpha^2(X - \mu)^2)$$
$$= \alpha^2 \text{Var}(X) \qquad\qquad \text{by Theorem 5.5(b) again.}$$

$\square$

Note that, except in a special case, as we will see in the next section, there is no nice general formula to express $\text{Var}(X + Y)$ solely in terms of $\text{Var}(X)$ and $\text{Var}(Y)$ (see also Exercise 5.8).

Theorem 5.6(a) is particularly convenient for practical calculations of $\text{Var}(X)$, as we see in the following.

Example 5.8 Find $\text{Var}(X)$ for the sum of scores on two fair dice.

Solution Using the data of Example 5.2, we find that

$$\mathbb{E}(X^2) = 4 \times \frac{1}{36} + 9 \times \frac{1}{18} + 16 \times \frac{1}{12} + 25 \times \frac{1}{9} + 36 \times \frac{5}{36} + 49 \times \frac{1}{6}$$
$$+ 64 \times \frac{5}{36} + 81 \times \frac{1}{9} + 100 \times \frac{1}{12} + 121 \times \frac{1}{18} + 144 \times \frac{1}{36}$$
$$= 54.833 \text{ (to 3 d.p.s)}.$$

From Example 5.6, we have $\mathbb{E}(X)^2 = 7^2 = 49$; hence by Theorem 5.6(a)

$$\text{Var}(X) = 54.833 - 49 = 5.833.$$

Example 5.9 Find the variance of:

(a) a Bernoulli random variable,
(b) a uniform random variable,
(c) the certain variable $X = x_k$.

Solution

(a) $\mathbb{E}(X^2) = (1)^2 p + (0)^2(1 - p) = p$, hence by Theorem 5.6(a) and the result of Example 5.7(a)

$$\text{Var}(X) = p - p^2 = p(1 - p).$$

(b) By direct evaluation in (5.5), we obtain

$$\sigma^2 = \frac{1}{n} \sum_{j=1}^{n} (x_j - \mu)^2.$$

This formula will be familiar to students of statistics as the 'population variance'.

(c) Again by (5.5) and the result of Example 5.7(c), we find

$$\sigma^2 = \sum_{j=1}^{n} (x_j - x_k)^2 \delta_{x_k} = 0.$$

The quantities $\mathbb{E}(X^n)$, where $n \in \mathbb{N}$, are called the *moments* of the random variable X. We have already seen that, when $n = 1$, we obtain the mean, and the first two moments determine the variance. Further properties of the moments will be investigated in the exercises.

5.4 Covariance and correlation

Suppose that X and Y are two random variables with means μ_X and μ_Y respectively. We say that they are *linearly related* if we can find constants m and c such that $Y = mX + c$, so for each $y_k (1 \leq k \leq m)$, we can find $x_j (1 \leq j \leq n)$ such that $y_k = mx_j + c$. If $m = n$ and we have $y_j = mx_j + c(1 \leq j \leq n)$, then each of the points $(x_1, y_1), (x_2, y_2), \ldots, (x_n, y_n)$ lies on a straight line as is shown in Fig. 5.3.

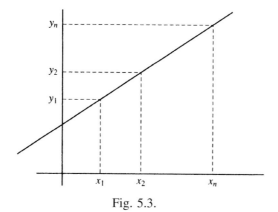

Fig. 5.3.

As Theorem 5.10 below will demonstrate, a quantity that enables us to measure how 'close' X and Y are to being linearly related is the *covariance* $\mathrm{Cov}(X, Y)$. This is defined by

$$\mathrm{Cov}(X, Y) = \mathbb{E}((X - \mu_X)(Y - \mu_Y)).$$

Note that $\mathrm{Cov}(X, Y) = \mathrm{Cov}(Y, X)$ and that $\mathrm{Cov}(X, X) = \mathrm{Var}(X)$. Furthermore (using the result of Exercise 5.8(b)), if X and Y are linearly related, then $\mathrm{Cov}(X, Y) = m \, \mathrm{Var}(X)$. Covariances are best calculated using the result of Exercise 5.8(a).

In practice, $\text{Cov}(X, Y)$ is measured in the product of the units of X and those of Y; however, the strength of the relationship between X and Y should be measured by a dimensionless number. For this reason, we define the *correlation coefficient* $\rho(X, Y)$ between X and Y by

$$\rho(X, Y) = \frac{\text{Cov}(X, Y)}{\sigma_X \sigma_Y},$$

where σ_X and σ_Y are the standard deviations of X and Y respectively. If $\rho(X, Y) = 0$, we say that X and Y are *uncorrelated*.

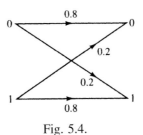

Fig. 5.4.

Example 5.10 Consider the binary symmetric channel shown in Fig. 5.4, where the input is described by a symmetric Bernoulli random variable X (so 0 and 1 are both emitted into the channel with probability $\frac{1}{2}$). Hence, by Exercise 4.15, the random variable Y describing the output is also symmetric Bernoulli. The error probability is given by $\varepsilon = 0.2$. Find the correlation $\rho(X, Y)$ between input and output.

Solution By Exercises 5.7(a) and 5.9(a), we find

$$\mu_X = \mu_Y = 0.5 \quad \text{and} \quad \sigma_X = \sigma_Y = 0.5.$$

Calculating joint probabilities by means of (4.1) yields

$$p_{00} = p_{11} = 0.4 \quad \text{and} \quad p_{01} = p_{10} = 0.1.$$

Hence, using the result of Exercise 5.8(a), we obtain

$$\text{Cov}(X, Y) = \mathbb{E}(XY) - \mathbb{E}(X)\mathbb{E}(Y)$$
$$= 0.4 - (0.5)^2 = 0.15,$$

so

$$\rho(X, Y) = \frac{0.15}{(0.5)(0.5)} = 0.6.$$

To explore further properties of $\text{Cov}(X, Y)$ and $\rho(X, Y)$ we need the following result, which is a vital tool in modern probability theory.

Theorem 5.7 *(the Cauchy–Schwarz inequality)*

$$\mathbb{E}(XY)^2 \le \mathbb{E}(X)^2\mathbb{E}(Y)^2.$$

Proof Consider the non-negative random variable $(X - tY)^2$ where $c \in \mathbb{R}$; then by Theorem 5.5(c) $\mathbb{E}((X - tY)^2) \geq 0$. Expanding the quadratic and using the result of Exercise 5.6 yields

$$t^2 \mathbb{E}(Y)^2 - 2t \mathbb{E}(XY) + \mathbb{E}(X)^2 \geq 0.$$

Now regarding the left-hand side of this expression as a quadratic function of t, the required result follows from the fact that $at^2 + bt + ct \geq 0$ for all t if and only if $b^2 \leq 4ac$. □

Corollary 5.8

(a) $|\text{Cov}(X, Y)| \leq \sigma_X \sigma_Y$.
(b) $-1 \leq \rho(X, Y) \leq 1$.

Proof

(a) Replace X by $X - \mu_X$ and Y by $Y - \mu_Y$ in the result of Theorem 5.7.
(b) Divide both sides of the result of (a) by $\sigma_X \sigma_Y$.

□

The final part of this section is a little bit harder than usual and may be omitted at first reading.

We begin with a definition. A pair of random variables (X, Y) is said to be *non-degenerate* if the joint probability $p_{jk} > 0$ for all $1 \leq j \leq n, 1 \leq k \leq m$ (i.e. none of the events $(X = x_j) \cap (Y = y_k)$ is impossible). The key result which we will need about pairs of non-degenerate random variables is the following technical lemma.

Lemma 5.9 *If (X,Y) is non-degenerate, then*

$$\mathbb{E}((X \pm Y)^2) = 0 \qquad \textit{if and only if } X \pm Y = 0.$$

Proof Immediate from the definition of non-degeneracy since

$$\mathbb{E}((X \pm Y)^2) = \sum_{j=1}^{n} \sum_{k=1}^{m} (x_j \pm y_k)^2 p_{jk}.$$

□

With the aid of Lemma 5.9 we can probe the relationship between the behaviour of $\rho(X, Y)$ and the concept of linearly related random variables.

Theorem 5.10 *Suppose that (X,Y) is non-degenerate, then:*

(a) $\mathbb{E}(XY)^2 = \mathbb{E}(X^2)\mathbb{E}(Y^2)$ *if and only if* $Y = mX$ *for some constant m,*
(b) $|\text{Cov}(X, Y)| = \sigma_X \sigma_Y$ *if and only if* X *and* Y *are linearly related,*
(c) $\rho(X, Y) = \pm 1$ *if and only if* X *and* Y *are linearly related.*

Proof

(a) The fact that $\mathbb{E}(XY)^2 = \mathbb{E}(X^2)\mathbb{E}(Y^2)$ when $Y = mX$ is left to Exercise 5.10(a). Conversely, suppose that $\mathbb{E}(XY)^2 = \mathbb{E}(X^2)\mathbb{E}(Y^2)$. Define two new random variables Z_1 and Z_2 by

$$Z_1 = \frac{X}{\mathbb{E}(X^2)^{1/2}} \quad \text{and} \quad Z_2 = \frac{Y}{\mathbb{E}(Y^2)^{1/2}}$$

then

$$\mathbb{E}(Z_1^2) = \mathbb{E}(Z_2^2) = 1 \qquad\qquad \text{by Theorem 5.5(b),}$$

and, since $\mathbb{E}(XY) = \pm\mathbb{E}(X^2)^{1/2}\mathbb{E}(Y^2)^{1/2}$, we find that

$$\mathbb{E}(Z_1 Z_2) = \pm 1 = \pm\mathbb{E}(Z_1^2) = \pm\mathbb{E}(Z_2^2).$$

Now use Theorem 5.5(a), to obtain

$$\mathbb{E}(Z_1 Z_2 \pm Z_1^2) = \mathbb{E}(Z_1 Z_2 \pm Z_2^2) = 0.$$

Adding these equations yields

$$\mathbb{E}((Z_1 \pm Z_2)^2) = 0.$$

Hence by Lemma 5.9, $Z_1 \pm Z_2 = 0$, thus

$$Y = mX \text{ with } m = \pm\left(\frac{\mathbb{E}(Y^2)}{\mathbb{E}(X^2)}\right)^{1/2}$$

(b) The fact that $|\text{Cov}(X, Y)| = \sigma_X \sigma_Y$ when X and Y are linearly related is again left to Exercise 5.10(b). To prove the converse replace X by $X - \mu_x$ and Y by $Y - \mu_Y$ in (a) to obtain the result with $m = \pm\frac{\sigma_Y}{\sigma_X}$ (depending on the sign of $\text{Cov}(X, Y)$) and $c = \mu_Y - m\mu_X$.

(c) is immediate from (b).

$\qquad\qquad\qquad\qquad\qquad\qquad\qquad\qquad\qquad\qquad\qquad\qquad\qquad\qquad\square$

Exercise: Extend Theorem 5.10 to the more general case where (X, Y) are no longer assumed to be non-degenerate.

5.5 Independent random variables

Two random variables X and Y are said to be (probabilistically) *independent* if each of the events $(X = x_j)$ and $(Y = y_k)$ are independent in the sense of the definition given in Section 4.4, that is

$$P((X = x_j) \cap (Y = y_k)) = P(X = x_j)P(Y = y_k)$$

for all $1 \le j \le n, 1 \le k \le m$. Using the notation introduced at the end of Section 5.2, this becomes

$$p_{jk} = p_j q_k \qquad\qquad\qquad\qquad (5.7)$$

Example 5.11 A child's toy contains two windows in each of which a clown's face appears when a button is pressed. The clown may be either smiling or crying. The design is such that the probability of each expression in each window is $\frac{1}{2}$. Define two symmetric Bernoulli random variables $X_j (j = 1, 2)$ by

$$X_j = 1 \text{ if clown in window } j \text{ is smiling}$$

$$X_j = 0 \text{ if clown in window } j \text{ is crying} \quad (j = 1, 2)$$

(a) Assuming that there is no mechanism linking the expression on the face of clown 1 to that of clown 2, show that X_1 and X_2 are independent.
(b) If there is a mechanism such that clown 1 smiles whenever clown 2 cries (and vice versa), show that X_1 and X_2 are not independent.

Solution

(a) We have $p_{11} = p_{12} = p_{21} = p_{22} = \frac{1}{4}$ and $p_1 = p_2 = q_1 = q_2 = \frac{1}{2}$; thus (5.7) is easily verified.
(b) Now $p_{11} = p_{22} = 0$, while $p_{12} = p_{21} = \frac{1}{2}$, while the ps and q_ks remain as above; thus we see immediately that (5.7) fails in all cases.

The following result collects some useful facts about independent random variables.

Theorem 5.11 *If X and Y are independent, then*:

(a) $\mathbb{E}(XY) = \mathbb{E}(X)\mathbb{E}(Y)$,
(b) $\text{Cov}(X, Y) = \rho(X, Y) = 0$,
(c) $\text{Var}(X + Y) = \text{Var}(X) + \text{Var}(Y)$.

Proof

(a) $\mathbb{E}(XY) = \sum_{j=1}^{n} \sum_{k=1}^{m} x_j y_k p_{jk}$

$\qquad = \sum_{j=1}^{n} \sum_{k=1}^{m} x_j y_k p_j q_k \qquad \text{by (5.7)}$

$\qquad = \left(\sum_{j=1}^{n} x_j p_j \right) \left(\sum_{k=1}^{m} y_k q_k \right)$

$\qquad = \mathbb{E}(X)\mathbb{E}(Y), \qquad\qquad \text{as required}$

(b) $\text{Cov}(X, Y) = \mathbb{E}(XY) - \mathbb{E}(X)\mathbb{E}(Y)$ by Exercises 5.8(a)

$\qquad\quad = 0 \qquad \text{by (a) above}$
$\quad$ Hence $\rho(X, Y) = 0$.
(c) This follows by (b) from Exercises 5.8(c). $\qquad\qquad\qquad\qquad\qquad$ □

In the case where X and Y are independent, we can obtain a useful prescription for the law of $X + Y$.

Suppose the law of X is p and the law of Y is q; then the law of $X + Y$ is called the *convolution* of p and q and is denoted as $p * q$ (see Exercises 3.18). For simplicity, we will take the range of X, $R_X = \{1, 2, \ldots, m\}$ and the range of Y, $R_Y = \{1, 2, \ldots, n\}$; then the range of $X + Y$, $R_{X+Y} = \{2, \ldots, m + n\}$. Let $k \in R_{X+Y}$; then we want

$$(p * q)(k) = p(X + Y = k).$$

Now the event $(X + Y = k)$ can occur in several different ways, that is it can occur as $(X = 1) \cap (Y = k-1)$ or $(X = 2) \cap (Y = k-2)$ or $\ldots$ or $(X = k-1) \cap (Y = 1)$, and these events form a partition of $(X + Y = k)$ so we have

$$(p * q)(k) = P[((X = 1) \cap (Y = k - 1)) \cup ((X = 2) \cap (Y = k - 2))$$

$$\cup \ldots \cup ((X = k - 1) \cap (Y = 1))]$$

$$= \sum_{j=1}^{k} P[(X = j) \cap (Y = k - j)] \qquad \text{by (P8)}$$

$$= \sum_{j=1}^{k} p_j q_{k-j} \qquad \text{by (5.7).} \qquad (5.8)$$

Formula (5.9) seems to lack symmetry between p and q; however, as $X + Y$ is the same random variable as $Y + X$, we would expect $(p * q) = (q * p)$. You can establish this result in Exercise 5.11(c) (see also Exercise 3.18).

Example 5.12 Return to the situation of Example 5.11(a). What is the probability that when the button is pressed there are (a) zero, (b) one, (c) two smiling clowns appearing in the windows?

Solution This question concerns the law of $X_1 + X_2$, thus we use (5.9):

(a) We require $P(X_1 + X_2 = 0) = (p * q)(0) = p_0 q_0 = \frac{1}{4}$.
(b) $P(X_1 + X_2 = 1) = p_0 q_1 + p_1 q_0 = \frac{1}{2}$.
(c) $P(X_1 + X_2 = 2) = p_1 q_1 = \frac{1}{4}$.

We generalise the definition of independent random variables to the case where there are more than two as follows. Let $X_1, X_2, \ldots, X_n$ be n random variables with ranges $R_j (1 \leq j \leq n)$. We say that they are independent if for every subset $\{i_1, i_2, \ldots, i_k\}$ of $\{1, 2, \ldots, n\}$ we have

$$P((X_{i_1} = x_{i_1}) \cap (X_{i_2} = x_{i_2}) \cap \ldots \cap (X_{i_k} = x_{i_k}))$$

$$= P(X_{i_1} = x_{i_1}) P(X_{i_2} = x_{i_2}) \ldots P(X_{i_k} = x_{i_k})$$

for all $x_{i_p} \in R_{i_p} (1 \leq p \leq k)$.

5.6 I.I.D. random variables

Two random variables X and Y are said to be *identically distributed* if they have the same range and

$$p_j = q_j \quad \text{for all } 1 \le j \le n.$$

If X and Y are both independent and identically distributed, then we say they are *i.i.d.* An example of such variables is the pair of symmetric Bernoulli random variables appearing in Example 5.10. In many situations in probability (as we will see below) we are interested in random variables $X_1, X_2, \ldots, X_n$, which are i.i.d. (i.e. the random variables are independent, as described at the end of Section 5.5, and all have the same range and the same law). In this case it is often useful to consider their sum

$$S(n) = X_1 + X_2 + \cdots + X_n.$$

We suppose that each $\mathbb{E}(X_j) = \mu$ and $\mathrm{Var}(X_j) = \sigma^2 (1 \le j \le n)$; then by Exercises 5.6 and 5.15, we have

$$\mathbb{E}(S(n)) = n\mu \qquad \text{and} \qquad \mathrm{Var}(S(n)) = n\sigma^2. \tag{5.9}$$

Example 5.13 [the simple random walk] Consider the above situation with each X_j defined by $R_j = \{-1, 1\}$ with common law $p(-1) = p(1) = \frac{1}{2}$ so each X_j has a similar form to that of a symmetric Bernoulli random variable. Consider a confused person who can't make up his/her mind where to go. S(h)e tosses a coin and if it lands on heads, s(h)e takes a step to the right; if it lands tails, s(h)e takes a step to the left. Suppose s(h)e repeats this procedure after every step, then $S(n)$ is the random variable which gives our friend's position after n steps. Note that $S(n)$ has range $\{-n, -n+2, \ldots, n-2, n\}$. Two possible paths which the person might take are sketched in Fig. 5.5. For example, the probability of returning to the starting point after two steps is

$$P(S(2) = 0) = P(((X_1 = 1) \cap (X_2 = -1)) \cup ((X_1 = -1) \cap (X_2 = 1)))$$

$$= P(X_1 = 1)P(X_2 = -1) + P(X_1 = -1)P(X_2 = 1) = \frac{1}{2}$$

where we have used (P3) and (5.7) (Fig. 5.6).

From (5.9), we find $\mathbb{E}(S(n)) = 0$ and $\mathrm{Var}(S(n)) = n$.

Application: elementary statistical inference

In statistics we are interested in gaining information about a *population* which for reasons of size, time or cost is not directly accessible to measurement, for example the population may consist of all the CDs manufactured by a company in a given week. We are interested in some quality of the members of the population which

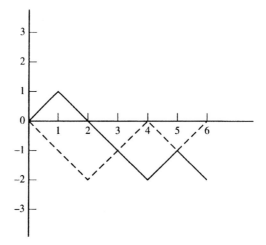

Fig. 5.5.

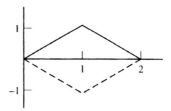

Fig. 5.6.

can be measured numerically (e.g. the number of defects on each CD). If we could measure all these numbers, we could calculate their mean and variance using the formulae given in Examples 5.7(b) and 5.9(b).

Statisticians attempt to learn about the population by studying a *sample* taken from it at random, for example we might take every 100th CD off the production line for inspection. Clearly, properties of the population will be reflected in properties of the sample. Suppose that we want to gain information about the population mean μ. If our sample is $\{x_1, x_2, \ldots, x_n\}$, we might calculate the sample mean

$$\overline{x} = \frac{1}{n} \sum_{j=1}^{n} x_j$$

and take this as an approximation to μ. However, there are many different samples of size n that we could take and we may have chosen a particularly poor one which gives a bad estimate. A more sophisticated technique is to try to consider all samples of size n simultaneously. We do this by using random variables. Let X_1 be the random variable whose values are all possible choices x_1 of the first member of the sample. In the simple model we are constructing, the range of X_1 can be the whole population, so $\mathbb{E}(X_1) = \mu$ and $\text{Var}(X_1) = \sigma^2$. Now let X_2

be the random variable whose values are all possible choices x_2 of the second member of the sample. In practice we are usually sampling without replacement, so the range of X_2 should be one less than the range of X_1, but statisticians are usually content to fudge this issue by arguing that if the population is suffi-ciently large, it is a 'reasonable' approximation to take X_1 and X_2 to be identically distributed. As the choice of value of X_1 should not affect that of X_2, we also assume that these random variables are independent. We continue in this way to obtain n i.i.d. random variables $X_1, X_2, \ldots, X_n$, where the range of X_j is the whole population considered as candidates to be the jth member of the sample ($1 \leq j \leq n$).

Now consider the random variable $\overline{X}(n)$ defined by

$$\overline{X}(n) = \frac{S(n)}{n} = \frac{1}{n} \sum_{j=1}^{n} X_j;$$

then the range of $\overline{X}(n)$ consists of the sample means obtained from all possible samples of size n taken from our population. By (5.9) and Theorem 5.6(b), we obtain

$$\mathbb{E}(\overline{X}) = \mu \text{ and } \mathrm{Var}(\overline{X}) = \frac{\sigma^2}{n}.$$

Further information about the law of $\overline{X}$ will be obtained in later chapters.

5.7 Binomial and Poisson random variables

Let $X_1, X_2, \ldots, X_n$ be i.i.d. Bernoulli random variables so that for each $1 \leq j \leq n$

$$P(X_j = 1) = p, \quad P(X_j = 0) = 1 - p.$$

The sum $S(n) = X_1 + X_2 + \cdots + X_n$ is called a *binomial random variable with parameters n and p*. Note that $S(n)$ is similar to the path of the simple random walk described in the last section, except, here, each X_j takes values in $\{0, 1\}$ rather than $\{-1, 1\}$. So the range of $S(n)$ is $\{0, 1, \ldots, n\}$.

Binomial random variables occur in situations where an operation which can be a 'success' (with probability p) or a 'failure' (with probability $1 - p$) is being repeated a given number of times. Then $S(n)$ counts the total number of successes; for example, suppose we have a sample of 100 CDs taken from a production line and an established criterion for quality which each CD passes with probability p; then the number of CDs which satisfy the quality standard is a binomial random variable with parameters 100 and p.

We now establish the probability law of $S(n)$. We could carry this out using repeated convolutions and, indeed, we have already dealt with the simple case where $n = 2$ and $p = \frac{1}{2}$ in this way in Example 5.12 above. We will, however, use a more direct method here

Lemma 5.12 *The probability law of the binomial random variable $S(n)$ is given by*

$$p(r) = \binom{n}{r} p^r (1 - p)^{n-r} \qquad \text{for } 1 \le r \le n. \tag{5.10}$$

Proof Recall that $S(n) = X_1 + X_2 + \cdots + X_n$, where each X_j is Bernoulli. We aim to calculate $p(r) = P(S(n) = r)$; hence, r of the X_js must take the value 1 and the other $(n - r)$ must take the value 0. As the X_js are independent, the probability of this event is $p^r (1 - p)^{n-r}$. However, there are $\binom{n}{r}$ mutually exclusive ways in which the event can be realised and the result follows by (P8). ☐

The probability law (5.10) is called the *binomial distribution*. By (5.9) and the results of Examples 5.7(a) and 5.9(a), we find immediately that

$$\mathbb{E}(S(n)) = np, \ \ \text{Var}(S(n)) = np(1 - p). \tag{5.11}$$

In practice we often 'forget' about the origins of binomial random variables as sums of i.i.d.s and call any random variable X binomial if it has the law (5.10). We will sometimes use the convenient notation $X \sim b(n, p)$ to mean 'X has a binomial distribution with parameters n and p'.

Example 5.14 Calculate the probability law and draw the probability histogram for $Y \sim b(6, 0.3)$. (See Fig. 5.7.)

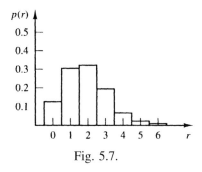

Fig. 5.7.

Solution Using (5.10), we obtain (to four significant figures)

$$p(0) = 0.1176, \ p(1) = 0.3025, \ p(2) = 0.3241, \ p(3) = 0.1852, \ p(4) = 0.0595$$
$$p(5) = 0.0102, \ p(6) = 0.0007.$$

In many applications $p(r)$ is the probability of r 'successes' in n experiences. Of course, it is also the probability of $(n - r)$ 'failures'. Indeed, if we introduce Y to be the random variable whose values are the total number of failures, then, since $\binom{n}{r} = \binom{n}{n-r}$ (see Chapter 2), we have $Y \sim b(n, 1 - p)$.

Example 5.15 An information source emits a six-digit message into a channel in binary code. Each digit is chosen independently of the others and is a one with probability 0.3. Calculate the probability that the message contains:
  (i) three 1s, (ii) between two and four 1s (inclusive), (iii) no less than two 0s.

Solution Let X be the number of 1s in the message, then $X \sim b(6, 0.3)$. Using the results of Example 5.14, we obtain:

(i) $p(3) = 0.1852$.
(ii) As the events $(X = 2)$, $(X = 3)$ and $(X = 4)$ are a partition of the event $(2 \leq X \leq 4)$, we have

$$P(2 \leq X \leq 4) = p(2) + p(3) + p(4) = 0.5688.$$

(iii) We require

$$P(X \leq 4) = p(0) + p(1) + p(2) + p(3) + p(4)$$

$$= 0.9889.$$

Having dealt with a finite number of i.i.d. Bernoulli random variables, it is natural (if you are a mathematician) to inquire about the behaviour of an infinite number of these. Of course, the passage to the infinite generally involves taking some kind of limit and this needs to be carried out with care.

We will take the limit of the probability law (5.10) as $n \to \infty$ and as $p \to 0$. In order to obtain a sensible answer, we will assume that n increases and p decreases in such a way that $\lambda = np$ remains fixed. We denote as $S_\lambda(\infty)$ the corresponding random variable, which is called a *Poisson random variable with mean* λ after the French mathematician S. Poisson, who first discussed it in the nineteenth century. To obtain the probability law of $S_\lambda(\infty)$, we substitute for p in (5.10) and (stressing the dependence of the binomial law on n and λ) so obtain

$$p_{n,\lambda}(r) = \frac{n(n-1)(n-2)\cdots(n-r+1)}{r!} \left(\frac{\lambda}{n}\right)^r \left(1 - \frac{\lambda}{n}\right)^{n-r}$$

$$= \frac{\lambda^r}{r!} \frac{n(n-1)(n-2)\cdots(n-r+1)}{n^r} \left(1 - \frac{\lambda}{n}\right)^{n-r}.$$

Now, as $n \to \infty$

$$\frac{n(n-1)(n-2)\ldots(n-r+1)}{n^r}$$

$$= 1 \left(1 - \frac{1}{n}\right) \left(1 - \frac{2}{n}\right) \cdots \left(1 - \frac{(r-1)}{n}\right) \to 1$$

and

$$\left(1 - \frac{\lambda}{n}\right)^{n-r} = \left(1 - \frac{\lambda}{n}\right)^{n}\left(1 - \frac{\lambda}{n}\right)^{-r} \to e^{-\lambda}$$

(where we have used the fact that $\lim_{n\to\infty}\left(1 - \frac{\lambda}{n}\right)^{n} = e^{-\lambda}$). Thus we see that $p_{n,\lambda}(r) \to p(r)$, where

$$p(r) = \frac{\lambda^{r}e^{-\lambda}}{r!} \tag{5.12}$$

for $r = 0, 1, 2, 3, \ldots$. Note that the range of $S_{\lambda}(\infty)$ is $\mathbb{N}$. If you are worried that the property of being a probability law may have been lost in the limit, you may confirm directly that $\sum_{r=0}^{\infty} p(r) = 1$ by using the well-known fact that

$$\sum_{r=0}^{\infty} \frac{\lambda^{r}}{r!} = e^{\lambda}.$$

The probability law (5.12) is called the *Poisson distribution with mean* λ.

If we take limits in (5.11), we find that

$$\mathbb{E}(S_{\lambda}(\infty)) = \lambda \quad \text{and} \quad \text{Var}(S_{\lambda}(\infty)) = \lambda. \tag{5.13}$$

To verify the second of these note that by (5.11),

$$\text{Var}(S(n)) = \lambda - \lambda p \to \lambda \quad \text{as } p \to 0.$$

If you prefer to obtain the results of (5.13) more directly, see Exercise 5.34(b).

In practice, as with the binomial case, any random variable X with probability law (5.12) is called Poisson. We introduce the notation $X \sim \pi(\lambda)$ to mean 'X has a Poisson distribution with mean λ'.

Example 5.16 Find the probability law and sketch the probability histogram for $X \sim \pi(1.3)$.

Solution Using (5.12), we find to four significant figures (Fig. 5.8)

$$p(0) = 0.2725, \quad p(1) = 0.3543, \quad p(2) = 0.2303,$$

$$p(3) = 0.0998, \quad p(4) = 0.0324, \quad p(5) = 0.0084,$$

$$p(6) = 0.0018, \quad p(7) = 0.0003, \quad p(r) = 0 \text{ (to five d.ps) for } r \leq 8.$$

In practice Poisson random variables arise when events occur in time or space in such a way that the average number of events occurring in a given time or space interval of fixed length is constant and events occurring in successive intervals are independent. The above example exhibits a hallmark of the Poisson distribution, namely that the probability of more than two events occurring in a fixed interval is 'small'. For this reason the Poisson distribution is often seen as a good model of 'rare' events. Examples of Poisson distributed random variables are the number

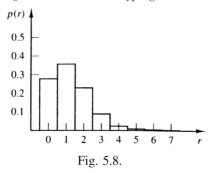

Fig. 5.8.

of currents in a slice of fruit cake, the number of misprints on the page of a book, the number of centenarians in a community, the number of stars in a volume of space and the number of atoms which decay within some time period in a piece of radioactive material.

Example 5.17 A message in binary code is being sent across a channel. Occasionally, there is a transmission error and an undecipherable splodge is sent instead of a 0 or a 1. The number of splodges transmitted per minute is found to be Poisson distributed with mean 1.3. What is the probability of more than two splodges per minute being sent?

Solution Using Lemma 5.3(i) and the results of Example 5.15 above, we find that

$$P(X > 2) = 1 - 0.2725 - 0.3543 - 0.2303 = 0.1429.$$

5.8 Geometric, negative binomial and hypergeometric random variables

In this section, we review some other useful types of random variable.

Geometric random variables

Let $X_1, X_2, \ldots$ be a sequence of i.i.d. Bernoulli random variables, each with parameter p, where $0 < p < 1$.

What is the probability (which we denote as $p(r)$) that the first $(r - 1)$ of these $X_1, X_2, \ldots, X_{r-1}$ all take the value 0 and X_r is 1?

By independence we have

$$p(r) = P((X_1 = 0) \cap (X_2 = 0) \cap \cdots \cap (X_{r-1} = 0) \cap (X_r = 1))$$
$$= (1 - p)^{r-1} p.$$

Now define a random variable Y taking values in $\mathbb{N}$ by

$$Y \text{ is the smallest value of } r \text{ for which } X_r = 1.$$

Y is called a *geometric random variable* and for each $r \in \mathbb{N}$ we have, by the above calculation,

$$P(Y = r) = p(r) = (1 - p)^{r-1} p; \qquad (5.14)$$

(5.14) is usually called the *geometric distribution with parameter p*.

In Exercise 5.28, you can verify for yourself that $\sum_{r=1}^{\infty} p(r) = 1$. It may be worth pointing out at this stage that the proof of this involves summing a geometric series using the well-known formula (for $0 < t < 1$)

$$\sum_{n=0}^{\infty} t^n = \frac{1}{1-t}. \qquad (5.15)$$

It is the connection with (5.15) that inspires the name of the random variable.

Geometric random variables are the simplest examples of random variables known as 'waiting times'. For example, suppose that a message in binary code is appearing on your terminal with a one-second gap between symbols. If we assume that symbols are generated in an i.i.d. manner, then the number of seconds we have to wait until the first one appears on the screen is a geometric random variable.

Example 5.18 A message in binary code received on a screen is such that waiting time until the first one is geometric with parameter $p = 0.3$. Find the probability that the first one appears as one of the first ten symbols.

Solution We want the cumulative probability

$$F(10) = P(Y \le 10).$$

This gives us a useful opportunity to develop a general formula. By (5.2) and (5.14) and using the formula for the sum of the first n terms of a geometric progression, we find that for arbitrary $0 < p < 1$ and $n \in \mathbb{N}$

$$F(n) = \sum_{r=1}^{n} p(1-p)^{r-1} = p\frac{1 - (1-p)^n}{1 - (1-p)}$$

$$= 1 - (1-p)^n.$$

Hence with $n = 10$ and $p = 0.3$ we have

$$F(10) = 1 - (0.7)^{10} = 0.97.$$

We will use a bit of trickery involving the well-known fact of elementary differentiation, that

$$\frac{d}{dp}((1-p)^r) = -r(1-p)^{r-1},$$

to calculate the mean of a general geometric random variable. By (5.4), (5.14) and (5.15) we have

$$\mathbb{E}(Y) = \sum_{r=1}^{\infty} r p (1-p)^{r-1} = -p \sum_{r=1}^{\infty} \frac{d}{dp}((1-p)^r)$$

$$= -p \frac{d}{dp} \sum_{r=0}^{\infty} (1-p)^r$$

$$= -p \frac{d}{dp}\left(\frac{1}{p}\right) = \frac{1}{p}. \tag{5.16}$$

You can use a similar technique to find $\mathbb{E}(Y^2)$, and hence $\mathrm{Var}(Y)$, in Exercise 5.29.

5.8.1 The negative binomial random variable

We will remain in the same context as above with our sequence $X_1, X_2, \ldots$ of i.i.d. Bernoulli random variables. We saw that the geometric random variable could be interpreted as a 'waiting time' until the first 1 is registered. Now we will consider a more general kind of waiting time, namely we fix $r \in \mathbb{N}$ and ask how long we have to wait (i.e. how many X_js have to be emitted) until we observe r 1s.

To be specific we define a random variable N called the *negative binomial random variable* with parameters r and p and range $\{r, r+1, r+2, \ldots\}$ by

$$N \text{ is the smallest value of } n \text{ for which } X_1 + X_2 + \cdots + X_n = r.$$

For example, if we preassign $r = 3$ and the sequence $010010100\ldots$ is generated, then we would have $n = 7$.

We now calculate the probability law of N. Let A be the event that $X_n = 1$ and B be the event that $(r-1)$ of the random variables $X_1, X_2, \ldots, X_{n-1}$ take the value 1. Of course, $P(A) = p$ and by (5.10) we have

$$P(B) = \binom{n-1}{r-1} p^{r-1}(1-p)^{n-r}.$$

We then have

$$P(N = n) = P(A \cap B)$$

$$= P(A)P(B) \qquad \text{by Exercise 5.16}$$

$$= \binom{n-1}{r-1} p^r (1-p)^{n-r}. \tag{5.17}$$

The law of (5.17) is called the *negative binomial distribution*. You should check that when $r = 1$ we recover the geometric distribution of the preceding section.

Example 5.19 We return to the context of Example 5.18. Find the probability that the third one is received as the seventh symbol.

Solution Taking $p = 0.3$, $r = 3$ and $n = 7$ in (5.17), we obtain

$$P(N = 7) = \binom{6}{2}(0.3)^3(0.7)^4 = 0.097.$$

As with the geometric distribution, the name 'negative binomial' arises from the series expansion used to check that it yields a genuine probability law; in fact we have

$$\sum_{n=r}^{\infty} \binom{n-1}{r-1} p^r (1-p)^{n-r}$$

$$= p^r \left(1 + r(1-p) + \frac{1}{2}r(r-1)(1-p)^2 \right.$$

$$\left. + \frac{1}{3!}r(r-1)(r-2)(1-p)^3 + \cdots \right)$$

$$= p^r \left(1 - r(-(1-p)) + \frac{1}{2}r(r-1)(1-p)^2 \right.$$

$$\left. - \frac{1}{3!}r(r-1)(r-2)(-(1-p))^3 + \cdots \right)$$

$$= p^r (1 - (1-p))^{-r} = 1, \text{ as required.}$$

5.8.2 The hypergeometric random variable

Let us suppose that we have a supply of n binary symbols (i.e. 0s and 1s), m of which take the value 0 (so that $n - m$ take the value 1). Suppose that we wish to form a 'codeword of length r', that is a sequence of r binary symbols out of our supply of n symbols and that this codeword is chosen at random.

We define a random variable

$$H = \text{number of 0s in the codeword}$$

so that H has range $\{0, 1, 2, \ldots, p\}$, where p is the minimum of m and r; H is called the *hypergeometric random variable* with parameters n, m and r.

We calculate the probability law of H, $P(H = k)$, where $0 \leq k \leq p$. By the basic principle of counting we see that the number of ways of choosing k 0s and $(r - k)$ 1s to make the required codeword is

$$\binom{m}{k}\binom{n-m}{r-k}.$$

As there are $\binom{n}{r}$ ways altogether of choosing the codeword, by the principle of symmetry we have

$$P(H = k) = \frac{\binom{m}{k}\binom{n-m}{r-k}}{\binom{n}{r}}. \tag{5.18}$$

The law of (5.18) is called the *hypergeometric distribution* (this name is inspired by the similarity in form to objects called hypergeometric functions).

Example 5.20 A bank of 18 binary symbols contains eight 0s. It is required to form a codeword of length 5. If such a codeword is chosen at random, what is the probability that it contains three 1s?

Solution We have $n = 18$, $m = 8$ and $r = 5$.

We want

$$P(H = 2) = \frac{\binom{8}{2}\binom{10}{3}}{\binom{18}{5}} = 0.39.$$

Although we have developed the hypergeometric distribution in the context of codewords, it is clearly applicable in any example where we are selecting from a large group containing two different kinds of object.

Exercises

5.1. A discrete random variable X has range $\{0,1,2,3,4,5\}$ and the first five terms of its cumulative distribution are

$$F(0) = 0, \ F(1) = \frac{1}{9}, \ F(2) = \frac{1}{6}, \ F(3) = \frac{1}{3} \text{ and } F(4) = \frac{1}{2}.$$

Find the probability law of X.

5.2. Three fair coins are tossed in succession. Find the probability law of the random variable 'total number of heads'.

5.3. If X is a random variable with range $\{x_1, x_2, \ldots, x_n\}$, show that:

(i) $P(X \geq x_r) = 1 - F(x_{r-1})$,
(ii) $P(x_r < X < x_s) = F(x_{s-1}) - F(x_r)$,
(iii) $P(x_r < X \leq x_s) = F(x_s) - F(x_r)$,
(iv) $P(x_r \leq X < x_s) = F(x_{s-1}) - F(x_{r-1})$.

5.4. Three tickets are chosen at random from a bag containing four red ones, five yellow ones, two blue ones and one red one. Use a combinatorial argument to obtain a formula for the probability law of the random variable: number of yellow tickets. Hence calculate each of these probabilities. [*Hint*: If you find this one difficult, you should try it again after reading Section 5.8.]

5.5. Find the mean, variance and standard deviation of the random variable of Example 5.1.

5.6. Use induction to show that

$$\mathbb{E}(X_1 + X_2 + \cdots + X_n) = \mathbb{E}(X_1) + \mathbb{E}(X_2) + \cdots + \mathbb{E}(X_n)$$

where $X_1, X_2, \ldots, X_n$ are arbitrary random variables.

5.7. Show that if c is constant, then

$$\text{Var}(X + c) = \text{Var}(X)$$

5.8. Show that, if X, Y, Y_1 and Y_2 are arbitrary random variables and α and β are real numbers:

(a) $\text{Cov}(X, Y) = \mathbb{E}(XY) - \mathbb{E}(X)\mathbb{E}(Y)$,
(b) $\text{Cov}(X, \alpha Y_1 + \beta Y_2) = \alpha \, \text{Cov}(X, Y_1) + \beta \, \text{Cov}(X, Y_2)$,
(c) $\text{Var}(X + Y) = \text{Var}(X) + 2 \, \text{Cov}(X, Y) + \text{Var}(Y)$.

5.9. Deduce from Exercise 5.8 above that

$$\text{Var}(X) + \text{Var}(Y) = \frac{1}{2}[\text{Var}(X + Y) + \text{Var}(X - Y)].$$

5.10. Show that:

(a) $\mathbb{E}(XY)^2 = \mathbb{E}(X^2)\mathbb{E}(Y^2)$ when $Y = mX$,
(b) $|\text{Cov}(X, Y)| = \sigma_X \sigma_Y$ when X and Y are linearly related.

5.11. If p, q and r are three probability laws, show that:

(i) $p * q = q * p$,
(ii)* $(p * q) * r = p * (q * r)$.

5.12. If X takes values $\{1, 2, \ldots, n\}$ with probability law p, find $\delta_1 * p$, where $1 \le 1 \le n$, and give an interpretation of your result in terms of random variables.

5.13. Show that X and Y are uncorrelated if and only if

$$\mathbb{E}(XY) = \mathbb{E}(X)\mathbb{E}(Y).$$

5.14. Let X and Y be independent random variables; show that

$$\text{Var}(X - Y) = \text{Var}(X + Y).$$

5.15. If X, Y and Z are independent random variables, show that $X + Y$ and Z are also independent. Hence, prove that if $X_1, X_2, \ldots, X_n$ are i.i.d. random variables

$$\text{Var}(S(n)) = \sum_{j=1}^{n} \text{Var}(X_j).$$

5.16. If X and Y are independent and E and F are arbitrary subsets in their ranges, show that

$$P((X \in E) \cap (Y \in F)) = P(X \in E)P(Y \in F).$$

5.17. Let $X_1, X_2, \ldots, X_n$ and $Y_1, Y_2, \ldots, Y_n$ be two sets of i.i.d. random variables such that X_i and Y_j are independent whenever $j \neq i$. Writing $S_n = X_1 + X_2 + \cdots + X_n$ and $T_n = Y_1 + Y_2 + \cdots + Y_n$, show that

$$\rho(S_n, T_n) = \frac{1}{n\sigma_X \sigma_Y} \sum_{j=1}^{n} \mathrm{Cov}(X_j, Y_j).$$

where σ_X (respectively σ_Y) is the common standard deviation of the X_is (respectively, the Y_js).

5.18. Consider a simple random walk $S(n) = X_1 + X_2 + \cdots + X_n$. Before each step, the random walker tosses a fair coin represented by the symmetric Bernoulli random variable Y_j taking values $\{0, 1\}$. For each $1 \leq j \leq n$, we are given the conditional probabilities

$$P_{(Y_j=1)}(X_j = -1) = 0.68, \quad P_{(Y_j=0)}(X_j = 1) = 0.79.$$

Find each $\rho(X_j, Y_j)$ and $\rho(S_n, T_n)$.

5.19. Find the probability that a simple random walk will be at the point (a) -1 after three steps, (b) 0 after four steps.

5.20. Find $\mathbb{E}(S(n))$ and $\mathrm{Var}(S(n))$ for a random walk in which each $P(X_j = 1) = p$ and $P(X_j = -1) = 1 - p$.

5.21. A random walk is the sum of n i.i.d. random variables for which

$$P(X_j = -1) = p, \quad P(X_j = 0) = 1 - p - q, \quad P(X_j = 1) = q.$$

Find $\mathbb{E}(S(n))$ and $\mathrm{Var}(S(n))$. Find also the probability that $P(S(2) = 0)$ when $p = 0.35$ and $q = 0.25$.

5.22. Let X have a uniform distribution with range $\{-2, -1, 1, 2\}$ and let $Y = X^2$ so that Y has range $\{1, 4\}$. Find the joint distribution of X and Y and hence deduce that X and Y are uncorrelated but not independent. (*Note*: This example appears on p. 222 of Feller's book.)

5.23. Fifteen items are selected from a production line, each of which has a probability of 0.3 of being defective. Use the binomial distribution to calculate the probability that (a) three items are defective, (b) more than four are defective, (c) more than 11 items are not defective, (d) less than five are defective, (d) between two and six (inclusive) are defective.

5.24. A binomial random variable has mean 10.4 and variance 3.64. Find the probability that it takes the value 7.

5.25. Show that if $X \sim b(n, 0.5)$, then

$$p(j) = p(n - j) \qquad \text{for } 1 \leq j \leq n.$$

Calculate the probability law when $n = 5$.

5.26. Show that the cumulative distribution $F(r)$ for a Poisson distribution of mean μ is

$$F(r) = e^{-\mu} \left(1 + \mu + \frac{\mu^2}{2!} + \cdots + \frac{\mu^r}{r!} \right).$$

5.27. A sample of a radioactive material emits α-particles at a rate of 0.7 per second. Assuming that these are emitted in accordance with a Poisson distribution, find the probability that in one second (i) exactly one is emitted, (ii) more than three are emitted, (iii) between one and four (inclusive) are emitted.

5.28. Verify that $\sum_{r=1}^{\infty} p(r) = 1$, where $p(r) = p(1-p)^{r-1}$ for $0 < p < 1$ is the probability law of a geometric random variable.

5.29.* Show that if Y is a geometric random variable with parameter p, then

$$\text{Var}(Y) = \frac{1-p}{p^2}.$$

[*Hint*: First calculate $\mathbb{E}(Y^2)$, write $r^2 = r(r-1) + r$ and imitate the argument for $\mathbb{E}(Y)$ using second derivatives this time.]

5.30. A binary message is transmitted over a channel at a rate of one symbol per second. Each symbol is independent of the preceding ones and has probability 0.6 of being a 1. Find the probability that:

(a) the first 1 is the second symbol to be received,
(b) the first 0 is the fourth symbol to be received,
(c) the message received is 01110,
(d) the second 1 is received as the fourth symbol.

5.31. Show that the mean of the negative binomial random variable N with parameters r and p is given by

$$\mathbb{E}(N) = \frac{r}{p}.$$

5.32. Convince yourself that N is the sum of r geometric random variables of parameter p and hence give an alternative verification of the result of the preceding question.

5.33. Use the hypergeometric distribution to find the probability that there are:

(i) two aces,
(ii) three aces,
(iii) four aces in a hand of poker.

5.34. If H is a hypergeometric random variable with parameters n, m and r, show that

$$P(H = k) = \frac{\binom{n-r}{m-k}\binom{r}{k}}{\binom{n}{m}}.$$

5.35. Use the identity $(x + y)^{m+n} = (x + y)^m (x + y)^n$ and the binomial theorem to show that

$$\binom{m+n}{k} = \sum_{j=0}^{k} \binom{m}{j}\binom{n}{k-j}.$$

Use this result and (5.8) to prove that if X and Y are independent with $X \sim b(n, p)$ and $Y \sim b(m, p)$, then $X + Y \sim b(n + m, p)$.

5.36. Show that if X and Y are independent with $X \sim \pi(\lambda)$ and $Y \sim \pi(\mu)$, then $X + Y \sim \pi(\lambda + \mu)$.

5.37. The *moment generating function* $M_X(t)$ of a random variable X is defined by

$$M_X(t) = \mathbb{E}(e^{tX})$$

for each $t \in \mathbb{R}$ so that

$$M_X(t) = \sum_{j=0}^{n} e^{tx_j} p_j$$

if X has finite range. Show that the nth moment of X

$$\mathbb{E}(X^n) = \frac{d^n}{dt^n} M_X(t)|_{t=0}.$$

5.38. (a) If $X \sim b(n, p)$, show that

$$M_X(t) = [pe^t + (1 - p)]^n.$$

(b) If $X \sim \pi(\lambda)$, show that

$$M_X(t) = \exp[\lambda(e^t - 1)].$$

Hence, using the result of Exercise 5.37, confirm that

$$\mathbb{E}(X) = \lambda = \text{Var}(X).$$

Find also $\mathbb{E}(X^3)$ and $\mathbb{E}(X^4)$ by this method.

5.39. (a) If X and Y are independent, show that

$$M_{X+Y}(t) = M_X(t)M_Y(t).$$

(b) If $X_1, X_2, \ldots, X_n$ are i.i.d. with common generating function $M(t)$, show that

$$M_{S(n)}(t) = (M(t))^n.$$

5.40. Let X be a random variable with range $\{x_1, x_2, \ldots, x_n\}$ and law $\{p_1, p_2, \ldots, p_n\}$ and let Y be a random variable with range $\{y_1, y_2, \ldots, y_m\}$ and law $\{q_1, q_2, \ldots, q_m\}$. We will use the notation

$$p_j(y_k) = P_{X=x_j}(Y = y_k), \quad (1 \le j \le n, 1 \le k \le m).$$

(i) Show that $\sum_{k=1}^{m} p_j(y_k) = 1$ for each $1 \le j \le n$.

For each $1 \le j \le n$, let $\tilde{Y}_j$ denote the random variable with range $\{y_1, y_2, \ldots, y_m\}$ and law $\{p_j(y_1), p_j(y_2), \ldots, p_j(y_m)\}$. For $1 \le j \le n$, the expectation value $\mathbb{E}(\tilde{Y}_j)$ is called the *conditional expectation of Y given $X = x_j$* and is sometimes denoted $\mathbb{E}_j(Y)$ or $\mathbb{E}_{X=x_j}(Y)$, so

$$\mathbb{E}_j(Y) = \sum_{k=1}^{m} y_k p_j(y_k).$$

We consider another random variable $\mathbb{E}_X(Y)$ called the *conditional expectation of Y given X* with range $\{\mathbb{E}_1(Y), \mathbb{E}_2(Y), \ldots, \mathbb{E}_n(Y)\}$ and law $\{p_1, p_2, \ldots, p_n\}$.

(ii) Show that $\mathbb{E}(\mathbb{E}_X(Y)) = \mathbb{E}(Y)$. [*Hint*: $\mathbb{E}(\mathbb{E}_X(Y)) = \sum_{j=1}^{n} \mathbb{E}_j(Y)p_j$.]

5.41.* Let X and Y be independent random variables each with range $\mathbb{N} \cup \{0\}$ and use the notation of the last question:

(a) Show that if $1 \leq k \leq n$, then

$$P_{X+Y=n}(X = k) = \frac{P(X = k)P(Y = n - k)}{P(X + Y = n)}.$$

(b) Now let X and Y be independent Poisson distributed random variables with means λ_1 and λ_2, respectively, and let $\tilde{X}_n$ be the random variable whose law is given by

$$P(\tilde{X}_n = k) = P_{X+Y=n}(X = k) \quad \text{for } 1 \leq k \leq n.$$

Show that $\tilde{X}_n$ has a binomial distribution and find its parameters.

(c) Find $\mathbb{E}_{X+Y=n}(X) = \mathbb{E}(\tilde{X}_n)$.

Further reading

Many of the books listed in preceding chapters can be consulted for more information about random variables. In particular Feller, Chung, Grimmett and Walsh, Ross and Rényi are highly recommended.

6

Information and entropy

6.1 What is information?

In this section we are going to try to quantify the notion of information. Before we do this, we should be aware that 'information' has a special meaning in probability theory, which is not the same as its use in ordinary language. For example, consider the following two statements:

(i) I will eat some food tomorrow.
(ii) The prime minister and leader of the opposition will dance naked in the street tomorrow.

If I ask which of these two statements conveys the most information, you will (I hope!) say that it is (ii). Your argument might be that (i) is practically a statement of the obvious (unless I am prone to fasting), whereas (ii) is extremely unlikely. To summarise:

(i) has very high probability and so conveys little information,
(ii) has very low probability and so conveys much information.
Clearly, then, quantity of information is closely related to the element of surprise.

Consider now the following 'statement':

(iii) XQWQ YK VZXPU VVBGXWQ.

Our immediate reaction to (iii) is that it is meaningless and hence conveys no information. However, from the point of view of English language structure we should be aware that (iii) has low probability (e.g. Q is a rarely occurring letter and is generally followed by U, (iii) contains no vowels) and so has a high surprise element.

The above discussion should indicate that the word 'information', as it occurs in everyday life, consists of two aspects, 'surprise' and 'meaning'. Of the above three examples, (i) has meaning but no surprise, (iii) has surprise but no meaning and only (ii) has both.

The mathematical theory of information which we are going to develop in this chapter is solely concerned with the 'surprise' aspect of information. There are two reasons for this. Firstly, information theory was originally developed within the context of communication engineering, where it was only the surprise factor that was of relevance. Secondly, 'meaning' has so far proved too difficult a concept to develop mathematically. The consequence of this is that we should be aware that 'information' in this chapter has the restricted technical meaning of 'measure of surprise'. Hence, statements such as (iii) above may well have a high information content, even though we consider them meaningless.

Let $(S, \mathcal{B}(S), P)$ be a probability space. In this chapter we will take $\#(S) = n$ and $\mathcal{B}(S) = \mathcal{P}(S)$. We would like to be able to measure the *information content* $I(E)$ of an event $E \in \mathcal{B}(S)$. From the above discussion, it seems clear that I should be a decreasing function of $P(E)$, the probability of E; that is, if $E, F \in \mathcal{B}(S)$ with $P(E) \leq P(F)$, then $I(E) \geq I(F)$. To gain further insight into the form of I, suppose that we draw a card at random from a pack of 52 and consider the following events:

   (i) the card is a heart (E_1),
   (ii) the card is a seven (E_2),
   (iii) the card is the seven of hearts ($E_1 \cap E_2$).

We have by the principle of symmetry, $P(E_1) = \frac{1}{4}$, $P(E_2) = \frac{1}{13}$, $P(E_1 \cap E_2) = \frac{1}{52}$. Note that E_1 and E_2 are independent events. From our above discussion, we have:

(a) $I(E_1 \cap E_2) \geq I(E_2) \geq I(E_1)$.

Our intuition tells that the amount of information $I(E_1 \cap E_2)$ that we get from learning (iii) is the sum of the information content of E_1 and E_2; that is, if E_1 and E_2 are independent, we have:

(b) $I(E_1 \cap E_2) = I(E_1) + I(E_2)$.

Together with (a) and (b) we will impose the commonsense condition that there is no such thing as negative information, that is:

(c) $I(E) \geq 0$ for all $E \in \mathcal{B}(S)$.

We now look for a candidate for a function which satisfies (a), (b) and (c). In fact, it can be shown that the only possibilities are of the form

$$I(E) = -K \, \log_a(P(E)) \qquad\qquad (6.1)$$

where a and K are positive constants. You should check using the laws of logarithms that (a), (b) and (c) above are all satisfied by (6.1), with the sole exception of events E for which $P(E) = 0$, in which case $I(E) = \infty$. Although (c) is violated in this case, we regard this as desirable – it indicates the non-feasibility of ever obtaining information about an impossible event. Equation (6.1) also has the desirable property that if the event E is certain, it contains no information as $\log_a(1) = 0$.

Note that since $\log_a(y) = \frac{\log_b(y)}{\log_b(a)}$, the choice of a is effectively a choice of the constant K and hence will only alter the units which $I(E)$ is measured in. Throughout this book we will make the standard choice $K = 1$ and $a = 2$. Hence, we define the information content of the event E by

$$I(E) = -\log_2(P(E)).$$
(6.2)

$K = 1$ is chosen for convenience. The choice of $a = 2$ is motivated by the following simple situation. Suppose that we have a symmetric Bernoulli random variable X taking values 0 and 1; then with the above convention we have

$$I(X = 0) = I(X = 1) = -\log_2\left(\frac{1}{2}\right) = 1.$$

The units of information content are *bits*. So we gain one bit of information when we choose between two equally likely alternatives.

As we will be using logarithms to base 2 extensively from now on, we will just write $\log_2(x) = \log(x)$. When directly calculating information content you should use the change of basis formula quoted above in the form $\log_2(x) = \frac{\ln(x)}{\ln(2)}$.

Example 6.1 A card is drawn at random from a pack of 52. What is the information content of the following events:

(i) the card is a heart,
(ii) the card is a seven,
(iii) the card is the seven of hearts.

Solution Using the probabilities calculated above, we have:

(i) $I(E_1) = -\log(\frac{1}{4}) = 2.00$ bits, as $4 = 2^2$.

(ii) $I(E_2) = -\log(\frac{1}{13}) = \frac{\ln(13)}{\ln(2)} = 3.70$ bits (to three significant figures).

(iii) Since E_1 and E_2 are independent, we have

$$I(E_1 \cap E_2) = I(E_1) + I(E_2) = 2 + 3.70 = 5.70 \text{ bits}.$$

We observe that the information content of an event depends only upon its probability. In the following, we will often be concerned with the events $(X = x_1)$,

$(X = x_2), \ldots, (X = x_n)$ arising from a discrete random variable X with range $\{x_1, x_2, \ldots, x_n\}$ and law $\{p_1, p_2, \ldots, p_n\}$. In this case we will write

$$I(p_j) = I(X = x_j) \quad (1 \le j \le n)$$

6.2 Entropy

Given a discrete random variable X, as described above, we cannot know for sure which of its values $x_1, x_2, \ldots, x_n$ will occur. Consequently, we don't know how much information $I(p_1), I(p_2), \ldots, I(p_n)$ we will be receiving, so that we may regard the information content of X itself as a random variable which we denote as $I(X)$. Clearly, it has range $\{I(p_1), I(p_2), \ldots, I(p_n)\}$. The mean of $I(X)$ is called its *entropy* and is denoted by $H(X)$, so that

$$H(X) = \mathbb{E}(I(X)) = -\sum_{j=1}^{n} p_j \log(p_j). \tag{6.3}$$

Note: In the case where $p_j = 0$, $p_j \log(p_j)$ is not well defined. We will define it to be 0 or, to be more specific, whenever you see the function $p \log(p)$ you should understand it to 'really mean' the function $\phi : [0, 1] \to [0, \infty)$, where

$$\phi(p) = p \log(p), \quad \text{when } p \ne 0,$$

$$\phi(0) = 0.$$

The use of the terminology 'entropy', which has its origins in thermodynamics and statistical physics, deserves some explanation. It was first introduced into information theory by its founder Claude Shannon (about whom more will be told at the end of the next chapter). When he first realised the importance of expression (6.3) in the theory, he consulted the great mathematician John von Neumann about a suitable name for it. Von Neumann's response (as reported by Myron Tribus) was as follows: 'You should call it 'entropy' and for two reasons: first, the function is already in use in thermodynamics under that name; second, and more importantly, most people don't know what entropy really is, and if you use the word 'entropy' in an argument you will win every time!' We will return to the connection between entropy and thermodynamics in Section 6.5 below. To gain some insight into the nature of entropy, we consider some examples.

Example 6.2 Find the entropy $H_b(p)$ of a Bernoulli random variable of parameter p.

Solution A simple application of (6.3) yields

$$H_b(p) = -p \log(p) - (1 - p) \log(1 - p). \tag{6.4}$$

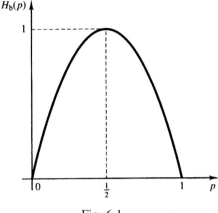

Fig. 6.1.

The graph of $H_b(p)$ against p is shown in Fig. 6.1. Note that $H_b(p)$ attains its maximum value of one bit when the random variable is symmetric, that is $p = \frac{1}{2}$ (see also Exercise 6.6).

Example 6.3 A coin is biased so that the probability of a head is (i) 0.95, (ii) 0.60, (iii) 0.5 (no bias). Calculate the entropy in each case.

Solution Using formula (6.4) yields the following:

(i) $H_b(0.95) = 0.286$ bits,
(ii) $H_b(0.60) = 0.971$ bits,
(iii) $H_b(0.5) = 1.000$ bit.

Let us consider Example 6.1 from the point of view of the person who has biased the coin and is now trying to make some money by gambling with it. How certain is s(h)e of winning at each toss? In (i) s(h)e is quite sure of winning and the entropy is low. In (ii) s(h)e is far less sure and the entropy is much higher. In (iii) s(h)e is in a state of maximum uncertainty and the entropy takes its maximum value. This leads us to the following conclusion:

Entropy is a measure of *uncertainty*.

In order to describe some of the general properties of entropy, we need the following very important inequality.

Lemma 6.1 $\ln(x) \le x - 1$ *with equality if and only if $x = 1$.*

Proof Figure 6.2 says it all, but those who aren't satisfied should try Exercise 6.4.

$\square$

Theorem 6.2 *Let X be a discrete random variable, then:*

(a) $H(X) \ge 0$ and $H(X) = 0$ if and only if X takes one of its values with certainty,
(b) $H(X) \le \log(n)$ with equality if and only if X is uniformly distributed.

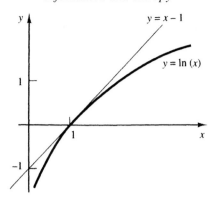

Fig. 6.2.

Proof

(a) Non-negativity of $H(X)$ is obvious from (6.3). Now suppose that $H(X) = 0$; then each $p_j \log(p_j) = 0$, hence we must have for some $k(1 \leq k \leq n)$ that $p_j = 0(j \neq k)$, $p_k = 1$.

(b) First suppose that $p_j > 0(1 \leq j \leq n)$. By (6.3) we have

$$H(X) - \log(n) = -\frac{1}{\ln(2)} \left(\sum_{j=1}^{n} p_j \ln(p_j) + \ln(n) \right)$$

$$= -\frac{1}{\ln(2)} \left(\sum_{j=1}^{n} p_j [\ln(p_j) + \ln(n)] \right) \qquad \text{by (5.1)}$$

$$= -\frac{1}{\ln(2)} \left(\sum_{j=1}^{n} p_j \ln(p_j n) \right) \qquad (\#)$$

$$= \frac{1}{\ln(2)} \left(\sum_{j=1}^{n} p_j \ln\left(\frac{1}{p_j n}\right) \right)$$

$$\leq \frac{1}{\ln(2)} \left(\sum_{j=1}^{n} p_j \left(\frac{1}{p_j n} - 1\right) \right) \qquad \text{by Lemma 6.1}$$

$$\leq \frac{1}{\ln(2)} \left(\sum_{j=1}^{n} \left(\frac{1}{n} - p_j\right) \right)$$

$$= 0 \qquad \text{by (5.1) again.}$$

By Lemma 6.1 we have equality if and only if $\frac{1}{p_j n} - 1 = 0$, that is each $p_j = \frac{1}{n}$, as is required.

Now suppose that $p_k = 0$ for some k; then returning to line (#) above, we have

$$-p_k \ln(p_k n) = 0 < \frac{1}{n} - p_k$$

and the result remains valid. □

The result of Theorem 6.2 should confirm our earlier intuition that entropy is a measure of uncertainty. Part (a) tells us that $H(X)$ is 0 precisely when we have zero uncertainty, and part (b) shows that entropy is a maximum precisely when we are maximally uncertain, that is when all options are equally likely. We will from now on write H_n to denote the entropy of a uniform distribution whose range has size n, so

$$H_n = \log(n).$$

We note that if $m \leq n$, $\log(m) \leq \log(n)$, so that $H_m \leq H_n$. Again, this confirms our intuition since we are more uncertain when we choose from a larger group of equally likely objects than when we choose from a smaller such group.

Part of the importance of the concept of entropy in probability theory is not just that it is a measure of uncertainty but that it is the only reasonable candidate to be such a measure. The 'uniqueness theorem' which establishes this result is a little more difficult than usual and so has been included, for those readers who are interested, in a separate section at the end of the chapter.

It may seem strange to some readers that we are using what is, by definition, the average information content of a random variable as a measure of its uncertainty. The key is to realise that uncertainty represents 'potential information' in the sense that when a random variable takes on a value we gain information and lose uncertainty.

6.3 Joint and conditional entropies; mutual information

Let X and Y be two random variables defined on the same probability space. We define their *joint entropy* $H(X, Y)$ to be

$$H(X, Y) = -\sum_{j=1}^{n}\sum_{k=1}^{m} p_{jk} \log(p_{jk}). \tag{6.5}$$

Clearly, $H(X, Y)$ is a measure of the combined uncertainty due to our ignorance of both X and Y. We note that $H(X, Y) = H(Y, X)$.

Example 6.4 Find the joint entropy of the random variables X_1 and X_2 defined in Example 5.11, on pages 76–7.

Solution

(a) We have $H(X_1, X_2) = -4 \times \frac{1}{4} \times \log(\frac{1}{4}) = $ two bits.
(b) $H(X_1, X_2) = -2 \times \frac{1}{2} \log(\frac{1}{2}) = $ one bit.

We note how the dependence between the random variables in (b) has led to a reduction in entropy. To explore the relationship between dependence and entropy more carefully we will need another entropy concept. First some notation: we will denote as $p_j(k)$ the conditional probability that $Y = k$ given that $X = j$. We then define the *conditional entropy of Y given that $X = j$, $H_j(Y)$* by

$$H_j(Y) = -\sum_{k=1}^{m} p_j(k) \log(p_j(k)) \tag{6.6}$$

where it is understood that $p_j(k) > 0$.

$H_j(Y)$ measures our uncertainty about Y when we know that the event $(X = x_j)$ has occurred.

Notes:

(i) From the point of view of Exercise 5.40, we have

$$H_j(Y) = H(\tilde{Y}_j).$$

(ii) If $p_j = 0$ so that $p_j(k)$ is undefined for all k, we define

$$H_j(Y) = H(Y).$$

Now consider the random variable $H.(Y)$, which has range $\{H_1(Y), H_2(Y), \ldots, H_n(Y)\}$ and probability law $\{p_1, p_2, \ldots, p_n\}$, so that $H.(Y)$ is a function of X. We define the *conditional entropy of Y given X, $H_X(Y)$* by

$$H_X(Y) = \mathbb{E}(H.(Y)) = \sum_{j=1}^{n} p_j H_j(Y), \tag{6.7}$$

so that $H_X(Y)$ is a measure of the uncertainty we still feel about Y after we know that X has occurred but don't know which value it has taken; ($H_X(Y)$ is sometimes called the *equivocation*).

Lemma 6.3

$$H_X(Y) = -\sum_{j=1}^{n}\sum_{k=1}^{m} p_{jk} \log(p_j(k)). \tag{6.8}$$

Proof Combine (6.6) and (6.7) to find that

$$H_X(Y) = -\sum_{j=1}^{n}\sum_{k=1}^{m} p_j p_j(k) \log(p_j(k)),$$

and the result follows from (4.1).   □

Lemma 6.4 *If X and Y are independent, then*

$$H_X(Y) = H(Y).$$

Proof Using the facts that $p_j(k) = q_k$ for $1 \le j \le n$, $1 \le k \le m$ when X and Y are independent, we see from (6.8) and (5.7) that

$$H_X(Y) = -\sum_{j=1}^{n}\sum_{k=1}^{m} p_j q_k \log(q_k) = H(Y) \qquad \text{by (5.1).}$$

$\square$

Example 6.5 A particle moves along the network shown above. The random variable X denotes its position after one second, for which there are two choices (labelled a and b) and the random variable Y is its position after two seconds, for which there are four choices (labelled 1, 2, 3 and 4). X is a symmetric Bernoulli random variable and we are given the following conditional probabilities (Fig. 6.3): for Y, $p_a(1) = \frac{2}{3}$, $p_a(2) = \frac{1}{3}$, $p_b(3) = p_b(4) = \frac{1}{2}$ (where $p_a(1) = p_{X=a}(Y = 1)$, etc.).

Calculate (a) $H_a(Y)$, (b) $H_b(Y)$, (c) $H_X(Y)$, (d) $H(X, Y)$.

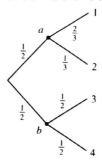

Fig. 6.3.

Solution

(a) Using (6.6)

$$H_a(Y) = -\frac{2}{3}\log\left(\frac{2}{3}\right) - \frac{1}{3}\log\left(\frac{1}{3}\right) = 0.918\,\text{bits.}$$

(b) Similarly

$$H_b(Y) = -2 \times \frac{1}{2}\log\left(\frac{1}{2}\right) = 1.000\,\text{bits.}$$

(c) $H_X(Y) = \frac{1}{2}H_a(Y) + \frac{1}{2}H_b(Y) = 0.959$ bits by (6.7).

(d) Using (4.1), we compute the joint probabilities

$$p(1, 1) = \frac{1}{3}, \; p(1, 2) = \frac{1}{6}, \; p(2, 3) = p(2, 4) = \frac{1}{4}.$$

Hence by (6.5), $H(X, Y) = 1.959$ bits.

Note that in the above example, we have

$$H(X, Y) - H_X(Y) = 1 = H(X).$$

More generally we have the following:

Theorem 6.5

$$H(X, Y) = H(X) + H_X(Y).$$

Proof Using (4.1) in (6.5) yields

$$H(X, Y) = -\sum_{j=1}^{n} \sum_{k=1}^{m} p_{jk} \log(p_j(k)p_j)$$

$$= -\sum_{j=1}^{n} \sum_{k=1}^{m} p_{jk} \log(p_j(k)) - \sum_{j=1}^{n} \sum_{k=1}^{m} p_{jk} \log(p_j)$$

and the result follows by Lemma 5.4(i).   □

Theorem 6.5 has the pleasant interpretation that the combined uncertainty in X and Y is the sum of that uncertainty which is totally due to X and that which is still due to Y once X has been accounted for. Note that, since $H(X, Y) = H(Y, X)$, we also have

$$H(X, Y) = H(Y) + H_Y(X).$$

Corollary 6.6 *If X and Y are independent*

$$H(X, Y) = H(X) + H(Y).$$

Proof Apply Lemma 6.4 to the result of Theorem 6.5.   □

Now $H_X(Y)$ is a measure of the information content of Y which is not contained in X; hence the information content of Y which is contained in X is $H(Y) - H_X(Y)$. This is called the *mutual information* of X and Y and is denoted $I(X, Y)$, so that

$$I(X, Y) = H(Y) - H_X(Y). \tag{6.9}$$

We collect some properties of mutual information in the following theorem:

Theorem 6.7

(a) $I(X, Y) = \sum_{j=1}^{n} \sum_{k=1}^{m} p_{jk} \log\left(\frac{p_{jk}}{p_j q_k}\right).$

(b) $I(X, Y) = I(Y, X).$

(c) *If X and Y are independent, then* $I(X, Y) = 0.$

Proof

(a) Using Lemma (5.4) (i) we find that

$$H(Y) = -\sum_{k=1}^{m} q_k \log(q_k) = -\sum_{j=1}^{n}\sum_{k=1}^{m} p_{jk} \log(q_k).$$

Hence, by (6.9) and (6.8),

$$I(X, Y) = -\sum_{j=1}^{n}\sum_{k=1}^{m} p_{jk} \log(q_k) + \sum_{j=1}^{n}\sum_{k=1}^{m} p_{jk} \log(p_j(k))$$

and the result follows when we write $p_j(k) = \frac{p_{jk}}{p_j}$.

(b) Immediate from (a).

(c) Follows from (6.9) and Lemma 6.4.

□

Note: Even if, say, $p_j = 0$ for some j, check via Lemma 5.4 that the formula in Theorem 6.7(a) is still meaningful.

From Theorem 6.7(b), we see that $I(X, Y)$ also measures the information about X contained in Y. We will gain more insight into this concept in the next chapter when we study information transmission.

Example 6.6 Calculate $I(X, Y)$ for the data of Example 6.4.

Solution In this case we have

$$q_1 = \frac{1}{3}, \ q_2 = \frac{1}{6} \quad \text{and} \quad q_3 = q_4 = \frac{1}{2}$$

hence

$$H(Y) = H(X, Y) = 1.959 \text{ bits}.$$

So using (6.9) and the solution of Example 6.4(c), we find

$$I(X, Y) = 1.959 - 0.959 = 1.000 \text{ bits}.$$

The interpretation of the solution of Example 6.5 is that Y contains all the information about X (i.e. 1 bit) or, alternatively, that none of the information contained in X is lost on the way to Y.

6.4 The maximum entropy principle

In Chapter 4, we introduced the symmetry principle for estimating unknown probabilities. Essentially, we argued that in a situation where we have no information about the events (i.e. we have maximum uncertainty) we should assume that the events are uniformly distributed. In Theorem 6.2 we have seen, however, that the uniform distribution occurs when we have maximum entropy and, furthermore (see Section 6.6 below), entropy is the unique measure of uncertainty. Hence we can rephrase the principle of symmetry as a 'principle of maximum entropy'. This would be a purely semantic operation except that our new principle is far more powerful than the old one in that it also gives us a mechanism for assigning probabilities when we have partial information about the events. To illustrate this principle in action we will describe an important example due to E. T. Jaynes, who first proposed this principle.

Let X be a random variable with range $\{x_1, x_2, \ldots, x_n\}$ and unknown law $\{p_1, p_2, \ldots, p_n\}$. Suppose that we have some information about X, namely that $\mathbb{E}(X) = E$, where E is some given constant. If E is different from the number given by Example 5.7(b), we know that X cannot have a uniform distribution. Which law should we associate to it? Using the technique of Lagrange multipliers (see Appendix 2 if this is not familiar to you) we maximise the entropy $H(X)$ subject to the two constraints

$$\text{(i) } \sum_{j=1}^{n} p_j = 1 \quad \text{and} \quad \text{(ii) } \sum_{j=1}^{n} x_j p_j = E.$$

Hence we must find the maximum value of the function of $(n+2)$ variables given by

$$L(p_1, p_2, \ldots, p_n; \lambda, \mu) = -\sum_{j=1}^{n} p_j \log(p_j) + \lambda \left(\sum_{j=1}^{n} p_j - 1 \right)$$

$$+ \mu \left(\sum_{j=1}^{n} x_j p_j - E \right),$$

where λ and μ are the Lagrange multipliers.

Differentiating yields the following $(n + 2)$ simultaneous equations in the unknowns

$$\frac{\partial L}{\partial p_j} = -\frac{1}{\ln(2)}(\ln(p_j) + 1) + \lambda + \mu x_j = 0 \ (1 \le j \le n)$$

and the two constraints (i) and (ii). Solving these for each p_j, we obtain n expressions of the type

$$p_j = \exp(\lambda' + \mu' x_j) \quad (1 \le j \le n)$$

where $\lambda' = \ln(2)\lambda - 1$ and $\mu' = \ln(2)\mu$.

From (i), we find that we must have $\lambda' = -\ln(Z(\mu'))$, where

$$Z(\mu') = \sum_{j=1}^{n} \exp(\mu' x_j) \tag{6.10}$$

thus we obtain for the entropy maximising probabilities

$$p_j = \frac{\exp(\mu' x_j)}{Z(\mu')} \tag{6.11}$$

for $1 \le j \le n$.

Now that λ has been eliminated it remains only to find the value of μ, but this is determined by (ii) above. Expression (6.10) is called the *partition function* and the probability law (6.11) is named the *Gibbs distribution* after the American physicist J. W. Gibbs (see Exercise 4.9). We will have more to say about the connection of (6.10) and (6.11) with physics in the next section. We conclude this one by giving a clear statement of the principle of maximum entropy.

6.4.1 Principle of maximum entropy

Given a random variable X with unknown law $p_1, p_2, \ldots, p_n$, we always choose the p_js so as to maximise the entropy $H(X)$ subject to any known constraints.

This principle gives a modern and far more powerful version of the principles of symmetry and insufficient reason, discussed within the context of the classical theory of probability in Section 4.2. In particular, it tells us that the Gibbs distribution (6.11) is the natural alternative to the uniform distribution when we are ignorant of all but the mean of our random variable.

6.5 Entropy, physics and life

Consider a liquid or gas inside some container. The fluid consists of a vast collection of particles in motion, all with different individual energies. We consider the random variable X whose values $x_1, x_2, \ldots, x_n$ are the possible energies of these particles, and apply the maximum entropy principle to find the law of X with the constraint that the average energy E is fixed. This is precisely the problem we faced in the last section and we know that the solution is given by (6.11). To make contact with known physical quantities, let T be the temperature of the fluid and define the inverse temperature parameter β by

$$\beta = \frac{1}{kT}$$

where k is a constant called *Boltzmann's constant* ($k = 1.38 \times 10^{-23}$ joules per kelvin). The Lagrange multipliers appearing in the preceding section are then given the following form:

$$\mu' = -\beta \text{ and } \lambda' = -\ln(Z(\mu)) = \beta F$$

where F is called the *Helmholtz free energy* (we will give a physical interpretation of F below). We thus obtain, for each $1 \le j \le n$

$$p_j = \exp \beta (F - x_i). \tag{6.12}$$

This distribution is well known in physics as that which describes the fluid in thermal equilibrium at temperature T.

Taking logarithms of both sides of (6.12), we obtain

$$\log(p_j) = \frac{\beta}{\ln(2)}(F - x_i).$$

Hence, on applying (4.1) and (5.4) in (6.3), we find

$$H(X) = -\sum_{j=1}^{n} p_j \log(p_j) = \frac{\beta}{\ln(2)}(E - F).$$

Now define the 'thermodynamic entropy' $S(X)$ by

$$S(X) = k \ln(2) H(X); \tag{6.13}$$

then we obtain the equation

$$F = E - T S(X). \tag{6.14}$$

Equation (6.14) is a well-known equation in statistical physics. Its interpretation is in terms of the law of conservation of energy. Recall that E is the average internal energy of the fluid, F is then the average energy of the fluid which is free to do work, while $T S(X)$ is the (heat) energy which maintains the fluid in equilibrium at temperature T. We remark that here we have obtained (6.14) as a simple consequence of the principle of maximum entropy.

The physicist Clausius originally introduced the concept of entropy into thermodynamics. He considered a small quantity of heat energy dQ absorbed by a system at temperature T and defined the entropy change dS by the formula

$$dS = \frac{dQ}{T}.$$

Now as heat can only flow from hot to cold bodies (and never vice versa), the only entropy changes that are possible are when a body loses heat at temperature T_1, which is then absorbed by another body at temperature T_2, where $T_2 \le T_1$. The corresponding entropy change is

$$-\frac{dQ}{T_1} + \frac{dQ}{T_2} \ge 0.$$

These considerations led Clausius, in 1865, to postulate the second law of thermodynamics, namely that the entropy of a closed system can never decrease. Indeed, as a closed system is, by definition, isolated from any interaction with the world outside itself, both observational evidence and the above considerations tell us that such a system should maximise its entropy and attain the Gibbs distribution (6.12) where it is in thermal equilibrium.

Entropy in physics is often described as a measure of 'disorder'. To understand why, one should appreciate that a fluid in equilibrium is in a changeless and uniform state. It is disordered in the sense that it is highly unlikely that the fluid will organise itself to leap out of the container and go and turn on the nearest television set! Such behaviour requires a high degree of order and would, in fact, correspond to entropy increase rather than decrease. Since such ordered, organised behaviour is the hallmark of living systems, the physicist E. Schrödinger, in his famous book *What is Life?* introduced the concept of 'negative entropy' $(-S)$ and argued that it is a characteristic of living things to absorb such negative entropy from their environment.

In this chapter, we have seen how the notion of entropy describes information, uncertainty and disorder. It is clearly a remarkable concept and it is worth closing this section by quoting the words of the astronomer A. Eddington from his book *The Nature of the Physical World* (written before the discovery of information theory).

Suppose that we were asked to arrange the following in two categories – distance, mass, electric force, entropy, beauty, melody.

I think there are the strongest possible grounds for placing entropy alongside beauty and melody, and not with the first three. Entropy is only found when the parts are viewed in association, and it is by viewing and hearing the parts in association that beauty and melody are discerned. All three are features of arrangement. It is a pregnant thought that one of these three associates should be able to figure as a commonplace quantity of science. The reason why this stranger can pass itself off among the aborigines of the physical world is that it can speak their language, viz. the language of arithmetic.

6.6 The uniqueness of entropy $(*)$

In this section, which is at a higher level of mathematical sophistication than the rest of this chapter, we will give a proof that entropy is the unique measure of uncertainty. The proof is based on that originally given by C. Shannon and then refined by A. I. Khintchin. It is not a prerequisite for any other part of this book and readers who find it too hard are encouraged to skip it.

Let X be a random variable with law $\{p_1, p_2, \ldots, p_n\}$. We say that a real valued function $U(X)$ (which we will sometimes write, where appropriate, as $U(p_1, p_2, \ldots, p_n)$) is a *measure of uncertainty* if it satisfies the following conditions:

(i) $U(X)$ is a maximum when X has a uniform distribution.

(ii) If Y is another random variable, then

$$U(X, Y) = U_X(Y) + U(X).$$

(iii) $U(p_1, p_2, \ldots, p_n, 0) = U(p_1, p_2, \ldots, p_n)$.

(iv) $U(p_1, p_2, \ldots, p_n)$ is continuous in all its arguments.

Before we present our main result, we comment on the definition above. We have derived (i) and (ii) already as properties of the entropy and argued as to why they are natural properties for a measure of uncertainty to possess. Item (iii) simply states that the uncertainty should not change when we also take into consideration the impossible event, and (iv) is a useful technical property. We also need to make some comments on the meaning of (ii). Given two random variables X and Y, we define $U_j(Y)$, the uncertainty of Y given that $X = x_j$, by

$$U_j(Y) = U(p_j(1), p_j(2), \ldots, p_j(m))$$

where the $p_j(k)$s are the usual conditional probabilities. We then define

$$U_X(Y) = \sum_{j=1}^{n} p_j U_j(Y).$$

It is not difficult to see that $U_X(Y) = U(Y)$ when X and Y are independent. Finally, the joint uncertainty $U(X, Y)$ is defined by

$$U(X, Y) = U(p_{11}, p_{12}, \ldots, p_{nm})$$

where the p_{ij}s are the joint probabilities for X and Y.

We are now ready to present the uniqueness theorem.

Theorem 6.8 $U(X)$ *is a measure of uncertainty if and only if*

$$U(X) = K H(X)$$

where $K \geq 0$ is a constant.

Proof Define $A(n) = U\left(\frac{1}{n}, \frac{1}{n}, \ldots, \frac{1}{n}\right)$. We will split our proof into three parts:

(a) In this part we show that $A(n) = K \log(n)$, thus establishing the theorem in the case where X is uniformly distributed. By (iii) and (i) we have

$$A(n) = U\left(\frac{1}{n}, \frac{1}{n}, \ldots, \frac{1}{n}, 0\right) \leq A(n + 1).$$

So A is a non-decreasing function of n.

Now let $X_1, X_2, \ldots, X_m$ be i.i.d. uniformly distributed random variables, each with r values in its range, so that each $U(X_j) = A(r)$, $1 \leq j \leq m$, then by (ii) we have

$$U(X_1, X_2) = U(X_1) + U(X_2) = 2A(r),$$

and by induction

$$U(X_1, X_2, \ldots, X_m) = m A(r).$$

However, the random vector $X = (X_1, X_2, \ldots, X_m)$ has r^m equally likely outcomes and so

$$U(X_1, X_2, \ldots, X_m) = A(r^m).$$

So we have that

$$A(r^m) = mA(r).$$

This result would also hold if we used n i.i.d. random variables, each with range of sizes s, that is

$$A(s^n) = nA(s).$$

Now choose r, s, n arbitrarily and let m be such that

$$r^m \leq s^n \leq r^{m+1} \qquad\qquad\qquad \text{P(i)}$$

(e.g. $r = 2$, $s = 3$ and $n = 4$ force us to take $m = 6$).
Using the fact that A is a non-decreasing function, we obtain

$$A(r^m) \leq A(s^n) \leq A(r^{m+1}),$$

hence

$$mA(r) \leq nA(s) \leq (m+1)A(r),$$

that is

$$\frac{m}{n} \leq \frac{A(s)}{A(r)} \leq \frac{m}{n} + \frac{1}{n};$$

and so

$$\left| \frac{A(s)}{A(r)} - \frac{m}{n} \right| \leq \frac{1}{n}. \qquad\qquad\qquad \text{P(ii)}$$

Now take logs of both sides of P(i) to obtain

$$m \log(r) \leq n \log(s) \leq (m+1) \log(r)$$

from which, by a similar argument to that given above, we find

$$\left| \frac{\log(s)}{\log(r)} - \frac{m}{n} \right| \leq \frac{1}{n}. \qquad\qquad\qquad \text{P(iii)}$$

Now, using the triangle inequality that for any two real numbers a and b, $|a+b| \leq |a| + |b|$, we obtain

$$\left| \frac{A(s)}{A(r)} - \frac{\log(s)}{\log(r)} \right| = \left| \left(\frac{A(s)}{A(r)} - \frac{m}{n} \right) + \left(\frac{m}{n} - \frac{\log(s)}{\log(r)} \right) \right|$$

$$\leq \left| \frac{A(s)}{A(r)} - \frac{m}{n} \right| + \left| \frac{\log(s)}{\log(r)} - \frac{m}{n} \right|$$

$$\leq \frac{2}{n} \text{ by P(ii) and P(iii).}$$

Since n can be made as large as we like, we must have

$$\frac{A(s)}{A(r)} = \frac{\log(s)}{\log(r)}$$

from which we deduce that $A(s) = K \log(s)$. The fact that A is non-decreasing yields $K \geq 0$. So we have completed part (a) of the proof.

(b) We will now prove the theorem in the case that each p_j is a rational number; to this end, we put

$$p_j = \frac{m_j}{m}, \quad \text{where} \sum_{j=1}^{n} m_j = m.$$

Now introduce another random variable Y which has m values and which we divide into n groups as follows

$$y_{11}, y_{12}, \ldots, y_{1m_1}, y_{21}, y_{22}, \ldots, y_{2m_2}, \ldots, y_{n1}, y_{n2}, \ldots, y_{nm_n}.$$

The reason for the strange grouping is that we want to make Y dependent on X and we do this by defining the following conditional probabilities where we condition on the event $X = x_r$

$$P_r(Y = y_{rk}) = \frac{1}{m_r} \qquad \text{for } 1 \leq k \leq m_r$$

$$P_r(Y = y_{sk}) = 0 \qquad \text{for } 1 \leq k \leq m_s, \; s \neq r$$

for $1 \leq r \leq n$.
    Hence, we obtain $U_r(Y) = K \log(m_r)$ by (a) and thus

$$U_X(Y) = \sum_{r=1}^{n} p_r U_r(Y) = K \sum_{r=1}^{n} \frac{m_r}{m} \log(m_r).$$

Now the joint probabilities

$$P((X = x_r) \cap (Y = y_{sk})) = p_r P_r(Y = y_{sk}) = 0 \qquad \text{when } s \neq r$$

$$= \frac{m_r}{m} \times \frac{1}{m_r} = \frac{1}{m} \qquad \text{for each } 1 \leq k \leq m_r.$$

Hence by (a) again and P(iii) we deduce that

$$U(X, Y) = K \log(m).$$

Now by P(ii), we find that

$$U(X) = U(X, Y) - U_X(Y)$$

$$= K \log(m) - K \sum_{r=1}^{n} \frac{m_r}{m} \log(m_r)$$

$$= -K \sum_{r=1}^{n} \frac{m_r}{m} \log\left(\frac{m_r}{m}\right), \quad \text{as required}$$

where we have used the fact that $\sum_{r=1}^{n} \frac{m_r}{m} = 1$. We have now completed the proof of (b).

(c) We now let the probabilities be arbitrary real numbers so each p_j can be approximated by a sequence of rationals $p_j^{(N)}$, where each $p_j^{(N)}$ can be written in the form given in (b) above. Let $H^{(N)}(X)$ be the corresponding sequence of entropies and define

$$H(X) = -\sum_{j=1}^{n} p_j \log(p_j);$$

then we have that $H(X) = \lim_{N \to \infty} H^{(N)}(X)$.

However, by the continuity assumption (iv) (p.106) and the result of (b), we also have

$$U(X) = \lim_{n \to \infty} H^{(N)}(X),$$

so by uniqueness of the limit, we must have $U(X) = H(X)$ and we have completed our proof. □

Exercises

6.1. Three possible outcomes to an experiment occur with probabilities $0.1, 0.3$ and 0.6. Find the information associated to each event.

6.2. You are told that when a pair of dice were rolled the sum on the faces was (a) 2, (b) 7. How much information is there in the two messages?

6.3. A word in a code consists of five binary digits. Each digit is chosen independently of the others and the probability of any particular digit being a 1 is 0.6. Find the information associated with the following events: (a) at least three 1s, (b) at most four 1s, (c) exactly two 0s.

6.4. Using the facts that $1/t \geq 1$ for $0 < t \leq 1$ and $1/t < 1$ for $t > 1$, show by integration that

$$\ln(x) \leq x - 1 \quad \text{for } x \geq 0.$$

6.5. Find the entropy when X is:

(a) the number of heads when two fair coins are tossed,
(b) distributed according to a Poisson law with mean 0.5.

6.6. The entropy of a Bernoulli random variable is given by

$$H_b(p) = -p \log(p) - (1 - p) \log(1 - p).$$

(a) Show that $H_b(0) = H_b(1) = 0$.

(b) Show that the graph of $H_b(p)$ is symmetric about the line $p = \frac{1}{2}$, that is

$$H_b\left(\frac{1}{2} - q\right) = H_b\left(\frac{1}{2} + q\right) \quad \text{for } 0 \le q \le \frac{1}{2}.$$

(c) Use calculus to show that $H_b(p)$ has a maximum at $p = \frac{1}{2}$. Hence sketch the graph of $H_b(p)$.

6.7. Show that

$$2^{-H(X)} = p_1^{p_1} p_2^{p_2} \cdots p_n^{p_n}.$$

6.8. X and Y are Bernoulli random variables with the distribution of X determined by $p = \frac{1}{3}$. We are also given the conditional probabilities

$$P_{X=0}(Y = 0) = \frac{1}{4} \quad \text{and} \quad P_{X=1}(Y = 1) = \frac{3}{5}.$$

Calculate (a) $H(X, Y)$, (b) $H_X(Y)$, (c) $H(Y)$, (d) $H_Y(X)$, (e) $I(X, Y)$.

6.9. Show that if $\{p_1, \ldots, p_n\}$ and $\{q_1, \ldots, q_n\}$ are two sets of probabilities, then we have the Gibbs inequality

$$-\sum_{j=1}^{n} p_j \log(p_j) \le -\sum_{j=1}^{n} p_j \log(q_j)$$

with equality if and only if each $p_j = q_j (1 \le j \le n)$. [*Hint:* First assume each $p_j > 0$, consider $\sum_{j=1}^{n} p_j \log\left(\frac{q_j}{p_j}\right)$ and then use Lemma 6.1 and (5.1).]

6.10. Using Gibbs inequality (or otherwise) show that

$$H_X(Y) \le H(Y)$$

with equality if and only if X and Y are independent.

Hence, deduce that $I(X, Y) \ge 0$ with equality if and only if X and Y are independent.

6.11. Let X be a random variable with law $\{p_1, p_2, \ldots, p_n\}$ and Y a random variable with law $\{q_1, q_2, \ldots, q_{n-1}\}$, where each

$$q_j = \frac{p_{j+1}}{1 - p_1} (1 \le j \le n - 1).$$

Show that

$$H(X) = H_b(p_1) + (1 - p_1)H(Y)$$

and hence deduce Fano's inequality

$$H(X) \le H_b(p_1) + (1 - p_1) \log(n - 1).$$

6.12. A random variable takes n possible values and only p_1 is known. Use the maximum entropy principle to deduce expressions for $p_2, p_3, \ldots, p_n$ and comment on these.

6.13. Three particles have energies 1, 2 and 3 J, respectively, and their mean energy is 2.4 J:

   (a) Use the maximum entropy principle to find their probability distribution.

   (b) (For those interested in physics.) Find the equilibrium temperature of the system and comment on its value.

6.14. Let X and Y be two random variables whose ranges are of the same size. Define the information theoretic distance (or relative entropy) $D(X, Y)$ of Y from X by

$$D(X, Y) = \sum_{j=1}^{n} p_j \log \left(\frac{p_j}{q_j} \right).$$

   (a) Show that $D(X, Y) \geq 0$ with equality if and only if X and Y are identically distributed.

   (b) Show that if Y has a uniform distribution, then

$$D(X, Y) = \log(n) - H(X).$$

   (c) Let W be the random vector (X, Y), so that the law of W is the joint distribution of X and Y, and let Z be the random vector whose law is given by that which W would have if X and Y were independent. Show that

$$D(W, Z) = I(X, Y).$$

6.15. The principle of minimum relative entropy asserts that a suitable posterior random variable X is that for which the relative entropy $D(X, Y)$ is minimised, where Y is the prior random variable. Show that when Y is uniformly distributed this is the same as requiring X to have maximum entropy.

6.16.* If we make a probability law 'more uniform', it seems reasonable that its entropy should increase. Establish this formally as follows: for the random variable X with range of sizes n, two of the probabilities p_1 and p_2 where $p_2 > p_1$ are replaced by $p_1 + \varepsilon$ and $p_2 - \varepsilon$, respectively, where $0 < 2\varepsilon < p_2 - p_1$. Prove that $H(S)$ is increased.

Further reading

The basic concepts of information, entropy, conditional entropy, etc., can be found in any book on information theory. The granddaddy of all these books is the ground-breaking *The Mathematical Theory of Information* by C. Shannon and W. Weaver (University of Illinois Press, 1949), which comprises a reprint of Shannon's original paper together with a lucid introduction by Weaver. A deeper mathematical account can be found in A. I. Khinchin's *Mathematical Foundations of Information Theory* (Dover, 1957). A fascinating discussion of the basics of information theory can be found in *A Diary on Information Theory* by A. Renyi (Wiley, 1984), which is written as a diary kept by a fictitious student attending a course of lectures on the subject. Some more references will be given at the end of the next chapter.

There is a nice discussion of entropy (including the maximum entropy principle) in Hamming's book quoted at the end of Chapter 4. For more on the maximum entropy principle, consult the conference proceedings *The Maximum Entropy Formalism* (MIT Press, 1979), edited by R. D. Levine and M. Tribus. Of particular interest therein is the introductory article by Tribus (from which the quoted discussion between Shannon and von Neumann is taken), and a superb review by Jaynes. This latter article contains some interesting reflections on the physical concept of entropy. *Maximum Entropy Models in Science and Engineering* by J. N. Kapur (Wiley Eastern, 1989) contains a wealth of interesting applications, as does *Maximum Entropy Solutions to Scientific Problems* by R. M. Bevensee (New York: Prentice Hall, 1993). For a more standard treatment of physical entropy, see the book by Isihara mentioned at the end of Chapter 2.

Eddington's *The Nature of the Physical World* is in the 'Everyman Series' (published by J. Dent and Sons Ltd, 1935). The quote in the text can also be found in Weaver's introduction to Shannon and Weaver as cited above. Schrödinger's *What is Life* was originally published in 1944 and has recently (1992) been reissued by Cambridge University Press within their Canto series.

7

Communication

7.1 Transmission of information

In this chapter we will be trying to model the transmission of information across channels. We will begin with a very simple model, as is shown in Fig. 7.1, and then build further features into it as the chapter progresses.

The model consists of three components. A source of information, a channel across which the information is transmitted and a receiver to pick up the information at the other end. For example, the source might be a radio or TV transmitter, the receiver would then be a radio or TV and the channel the atmosphere through which the broadcast waves travel. Alternatively, the source might be a computer memory, the receiver a computer terminal and the channel the network of wires and processors which connects them. In all cases that we consider, the channel is subject to 'noise', that is uncontrollable random effects which have the undesirable effect of distorting the message leading to potential loss of information by the receiver.

The source is modelled by a random variable S whose values $\{a_1, a_2, \ldots, a_n\}$ are called the *source alphabet*. The law of S is $\{p_1, p_2, \ldots, p_n\}$. The fact that S is random allows us to include within our model the uncertainty of the sender concerning which message they are going to send. In this context, a message is a succession of symbols from S sent out one after the other. For example, to send a message in the English language we would take $n = 31$ to include 26 letters, a blank space to separate words, a comma, full-stop, question mark and exclamation mark.

The receiver R is another random variable with range $\{b_1, b_2, \ldots, b_m\}$, called the *receiver alphabet*, and law $\{q_1, q_2, \ldots, q_m\}$. Typically $m \geq n$ (see e.g. Exercises 4.16 and 4.17). Finally, we model the distorting effect of the channel on messages by the family of conditional probabilities $\{p_i(j); 1 \leq i \leq n, 1 \leq j \leq m\}$, where we have, as usual, written each $P_{S=a_i}(R = b_j)$ as $p_i(j)$. Clearly, for optimal successful transmission in, for example, the case $n = m$ where both alphabets are identical, we would like each $p_i(i)$ to be as close to 1 as possible.

```
┌──────────┐      ┌──────────┐      ┌──────────┐
│  SOURCE  ├──►   │ CHANNEL  ├──►   │ RECEIVER │
└──────────┘      └──────────┘      └──────────┘
```

Fig. 7.1.

In this chapter we will try to formulate our ideas in the context of the simplest possible model, namely the *binary symmetric channel*, which we studied earlier in Chapter 4. For convenience, we have again sketched the model in Fig. 7.2. However, readers may find it useful to refresh their memories by re-reading the appropriate pages in Section 4.3.

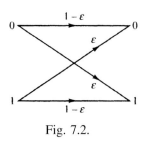

Fig. 7.2.

Note that the noise in the channel is entirely described by the variable ε in that $p_0(1) = p_1(0) = \varepsilon$, which, ideally, we would like to be as small as possible. The source and receiver alphabets are both the binary alphabet $\{0, 1\}$. Furthermore, as was shown in Section 4.3, the law of R is completely determined by knowledge of ε and the law of S.

A useful concept for modelling the flow of information across channels is the *mutual information* between two events E and F. This is not the same concept as the mutual information between two random variables which we discussed in the last section; however, as we will see below, they are related. We will define mutual information in the context discussed above where E is the event $(S = a_j)$ and F is the event $(R = b_k)$ and denote it by the symbol $I(a_j, b_k)$ so that, if $p_j > 0$, we define

$$I(a_j, b_k) = -\log(q_k) + \log(p_j(k)). \tag{7.1}$$

(If $p_j = 0$, we let $I(a_j, b_k) = 0$.)

As $-\log(q_k)$ is the information content of the event $(R = b_k)$ and $-\log(p_j(k))$ is the remaining information we would need to specify that $(R = b_k)$ given that $(S = a_j)$ was transmitted; we can interpret $I(a_j, b_k)$ as the information about $(R = b_k)$ which is contained in $(S = a_j)$. In other words, it measures the quantity of information transmitted across the channel. Notice that if there is no noise in the channel, we have

$$I(a_j, b_k) = -\log(q_k) = I(q_k)$$

so that if a_j is sent out with the intention of b_k being received, then this is precisely what happens. The following theorem collects some useful properties of mutual information (see Theorem 6.7).

Theorem 7.1 *For each* $1 \leq j \leq n$, $1 \leq k \leq m$, *we have:*

(i) $I(a_j, b_k) = \log \left(\frac{p_{jk}}{p_j q_k} \right)$.

(ii) $I(a_j, b_k) = -\log(p_j) + \log(q_k(j))$, *where* $q_k(j) = P_{R=b_k}(S = a_j)$.

(iii) $I(a_j, b_k) = I(b_k, a_j)$.

(iv) *If the events* $(S = a_i)$ *and* $(R = b_j)$ *are independent, then* $I(a_j, b_k) = 0$.
   Furthermore, we have that

(v) $I(S, R) = \sum_{j=1}^{n} \sum_{k=1}^{m} p_{jk} I(a_j, b_k)$.

Proof

(i) is deduced from $p_j(k) = \frac{p_{jk}}{p_j}$.

(ii) follows from (i) and the fact that each

$$q_k(j) = \frac{p_{jk}}{q_k}$$

(iii) and (iv) both follow immediately from (i).

(iv) follows from (i) and Theorem 6.7(a). □

Item (iii) has an interesting interpretation, namely that if the channel were inverted so that S became the receiver and R the source, then the information about a_j contained in b_j would be exactly the same as that about b_j contained in a_k when the channel was operating normally. Item (v) expresses the mutual information between the source and receiver random variables (as defined by (6.9)) as the average over all possible transmissions of individual symbols.

Example 7.1 Find the mutual informations $I(a_0, b_0)$, $I(a_0, b_1)$, $I(a_1, b_0)$ and $I(a_1, b_1)$ for a binary symmetric channel and comment on these expressions.

Solution Note that, here, $a_0, b_0 = 0$ and $a_1, b_1 = 1$.

Using (7.1) and the formulae for the law of R found in Section 4.3, we obtain

$$I(a_0, b_0) = -\log(1 - \varepsilon - p + 2\varepsilon p) + \log(1 - \varepsilon)$$

$$= \log \left(\frac{1 - \varepsilon}{1 - \varepsilon - p + 2\varepsilon p} \right)$$

Similar arguments yield

$$I(a_1, b_1) = \log\left(\frac{1 - \varepsilon}{\varepsilon + p - 2\varepsilon p}\right),$$

$$I(a_0, b_1) = \log\left(\frac{\varepsilon}{\varepsilon + p - 2\varepsilon p}\right),$$

$$I(a_1, b_0) = \log\left(\frac{\varepsilon}{1 - \varepsilon - p + 2\varepsilon p}\right).$$

Suppose now that we put $\varepsilon = 0$ so there is no noise in the channel; then we find $I(a_0, b_0) = I(a_0) = -\log(1 - p)$, $I(a_1, b_1) = I(a_1) = -\log(p)$ and $I(a_0, b_1) = I(a_1, b_0) = -\infty$, as we would expect (i.e. a perfect transmission of information). If we put $\varepsilon = 1$, we obtain similar results except that this time $I(a_0, b_1) = I(a_0)$ and $I(a_1, b_0) = I(a_1)$. Again there is perfect transmission of information except that all the information about a_0 is sent to b_1 and all the information about a_1 is sent to b_0. If we put $\varepsilon = \frac{1}{2}$, so the noise produces maximum uncertainty, we find that $I(a_0, b_0) = I(a_1, b_1) = I(a_0, b_1) = I(a_1, b_0) = 0$, that is no information is transmitted.

Suppose now that S is symmetric Bernoulli (in which case R is also – see Exercise 4.15). We then obtain from the above formulae

$$I(a_0, b_0) = I(a_1, b_1) = \log(2(1 - \varepsilon)) = 1 + \log(1 - \varepsilon),$$

$$I(a_0, b_1) = I(a_1, b_0) = \log(2\varepsilon) = 1 + \log(\varepsilon).$$

In the case where $0 < \varepsilon < \frac{1}{2}$, both $I(a_0, b_1)$ and $I(a_1, b_0)$ are negative, while $I(a_0, b_0)$ and $I(a_1, b_1)$ are less than 1; whereas if $\frac{1}{2} < \varepsilon < 1$, then both $I(a_0, b_0)$ and $I(a_1, b_1)$ are negative, while $I(a_0, b_1)$ and $I(a_1, b_0)$ are less than 1.

Negative mutual information is an indication that the effects of the noise are such that all information about the emitted symbol has been lost and hence the information transmitted is more likely to tell us about the characteristics of the noise rather than those of the signal.

We note that while mutual information between events may, as seen above, sometimes be negative, the mutual information $I(S, R)$ between source and receiver random variables can never be negative (see Exercises 6.10).

7.2 The channel capacity

In the preceding section, we have investigated the information transmission properties of channels from the point of view of their ability to transmit information between individual events. For the broader picture in which we consider the transmission properties of the channel as a whole, the key concept is the mutual information $I(S, R)$.

Suppose we fix the channel by choosing all the conditional properties $\{p_j(k), 1 \leq j \leq n, 1 \leq k \leq m\}$. We would like to ensure that the channel transmits the maximum possible amount of information about our message. The only variables we have left to alter are then the source probabilities $\{p_1, p_2, \ldots, p_n\}$.

We define the *channel capacity* C by

$$C = \max I(S, R)$$

where the maximum is taken over all possible probability laws of the random variable S. In practice, the best course is often to use (6.9) and apply

$$C = \max(H(R) - H_S(R)).$$

Example 7.2 Find the channel capacity of a binary symmetric channel.

Solution Here we just have to find the maximum with respect to the variable x which describes the Bernoulli distribution of R where $x = \varepsilon + p - 2\varepsilon p$ (see Section 4.3).

We have

$$H(R) = H_b(x) = -x \log(x) - (1 - x) \log(1 - x).$$

A standard calculation of the joint probabilities yields

$$p_{00} = (1 - p)(1 - \varepsilon), \quad p_{01} = (1 - p)\varepsilon,$$
$$p_{10} = p(1 - \varepsilon), \quad p_{11} = p\varepsilon.$$

Hence, by (6.8)

$$H_S(R) = -(1 - p)(1 - \varepsilon) \log(1 - \varepsilon) - (1 - p)\varepsilon \log(\varepsilon)$$
$$- p(1 - \varepsilon) \log(1 - \varepsilon) - p\varepsilon \log(\varepsilon)$$
$$= -(1 - \varepsilon) \log(1 - \varepsilon) - \varepsilon \log(\varepsilon) = H_b(\varepsilon).$$

Since $H_S(R)$ is clearly independent of p, we have

$$C = \max(H_b(x)) - H_b(\varepsilon)$$
$$= 1 - H_b(\varepsilon) \text{ by Theorem 6.2(b).}$$

In Exercise 7.2, you can show that C is in fact attained when $p = \frac{1}{2}$.

We note that C is a function of ε and sketch the relationship in Fig. 7.3.

Clearly, if we want to transmit information as near as possible to the channel capacity C, we need to adjust the input probabilities accordingly. At first sight this does not seem possible. If we want to send a message, then the probabilities

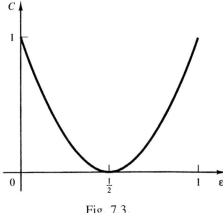

Fig. 7.3.

of the symbols that constitute that message are fixed; however, there is one way out – we can change the language in which the message is sent to one with 'better' probabilistic properties. This takes us to the subject of coding, which will occupy us in the next section.

7.3 Codes

The mathematical theory of codes has been heavily developed during recent years and has benefited from the application of sophisticated techniques from modern abstract algebra. Our goal here will be a limited one – simply to learn enough about codes to be able to advance our knowledge of information transmission. First some definitions. A *code alphabet C* is simply a set $\{c_1, c_2, \ldots, c_r\}$, the elements of which are called *code symbols*. A *codeword* is just a certain succession of code letters $c_{i_1} c_{i_2} \ldots c_{i_n}$. The number n occurring here is the *length* of the codeword. A *code* is the set of all possible codewords, and a *code message* is then a succession of one or more codewords. The process of coding a language is to map every symbol in the alphabet of that language into an appropriate codeword. A common example of a code for the English language is Morse code, which consists of just two symbols, · (dot) and – (dash). Much of our work in this chapter will involve the following:

Example 7.3 [binary code] This is the simplest code of all but one that is of great importance (especially for the operation of computers). We have

$$C = \{0, 1\}.$$

Examples of codewords are

$$x = 011010, \quad y = 1010, \quad z = 01;$$

x has length 6, *y* has length 4 and *z* has length 2. If we demand (say) that our code consists of all codewords of length 6 or less, then we should check that there are 126 possible codewords. A possible message is then

$$01 \quad 10001 \quad 101010 \quad 01$$

In practice, of course, there are no gaps in the message to tell you where one codeword ends and another begins and this can be a source of difficulty in decoding, as we will see below.

An important code for all living creatures is the following.

Example 7.4 [the genetic code] Inside the nucleus of each living cell can be found the giant molecule of DNA (deoxyribonucleic acid), which has the characteristic double helix shape. Simple organisms such as a bacteria have a single DNA helix in their nucleus. Higher organisms have many bundles of helices called chromosomes, for example onions have 16, human beings 42 and cattle 60 of these. Along each DNA strand are embedded certain molecular bases adenine (A), thymine (T), cytosine (C) and guanine (G). The bases occur in pairs on opposite strands of the helix and there are approximately 10^4 such pairs in the DNA of a bacteria and 10^9 in that of a mammal. These bases form the alphabet {A, T, C, G} for the genetic code. DNA directs protein synthesis in the cell by the following procedure. All proteins are built from 20 possible amino acids. By a complicated biochemical procedure DNA 'tells' the cell which amino acids to produce by sending out a message. The allowable codewords are all of length three. Clearly, there are $4^3 = 64$ possibilities available to make the 20 amino acids (and the one 'stop' command to cease manufacturing). Consequently, several different codewords may instruct for the manufacture of the same amino acid and the code has a high degree of 'redundancy', for example the codewords CAT and CAC both instruct for the production of the amino acid histidine.

We aim to understand the implications for information transmission of coding our source messages before they are transmitted across the channel, thus we now have the more elaborate model of a communication channel shown in Fig. 7.4.

When translating a source alphabet of size n using a code alphabet of size r, it is usual to take $r \leq n$ (e.g. $n = 26$ for English and $r = 2$ for Morse code). We will use examples based on the binary code with alphabet {0, 1} extensively as this is

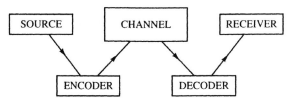

Fig. 7.4.

the natural code for storing information in computer systems. We now consider a number of desirable properties which we will require of our codes. Clearly:

 (i) Each symbol of the source alphabet should be mapped to a unique codeword.
(ii) The code should be *uniquely decodeable*, that is any finite sequence of codewords should correspond to one and only one message.

Example 7.5 We code the source alphabet $\{a_1, a_2, a_3, a_4, a_5\}$ into binary as follows:
$$a_1 \to 0, \ a_2 \to 1, \ a_3 \to 10, \ a_4 \to 01, \ a_5 = 00.$$

Is the code uniquely decodeable?

Solution It is not; for, consider the message

$$0001100110$$

This message has a number of different translations, for example $a_1 a_5 a_2 a_3 a_1 a_2 a_3$ and $a_1 a_1 a_4 a_2 a_5 a_2 a_3$ are two possibilities; you should be able to find many others.

 Uniquely decodeable codes are obviously highly desirable; however, an even better class of codes are those that have the property that a message sent in the code can be instantaneously read as each codeword appears without having to wait for the whole message to arrive. The concept we need to express this formally is that of a prefix: let $c_{i_1} c_{i_2} \ldots c_{i_n}$ be a codeword and suppose that for some $k < n$, $c_{i_1} c_{i_2} \ldots c_{i_k}$ is also a codeword; then $c_{i_1} c_{i_2} \ldots c_{i_k}$ is called a *prefix*, for example 010 is a prefix of 01011. A code for which no codeword is a prefix for any other is called *prefix-free* (or *instantaneous*).

Example 7.6 Consider the five-symbol alphabet of Example 7.5. Which (if any) of the following possible codes is prefix-free?

 (i) $a_1 = 00$, $a_2 = 01$, $a_3 = 110$, $a_4 = 010$, $a_5 = 01011$,
(ii) $a_1 = 0$, $a_2 = 10$, $a_3 = 110$, $a_4 = 1110$, $a_5 = 11110$.

Solution Clearly, (i) is not prefix-free as 01 is a prefix of 010, which is itself a prefix of 01011 (you should, however, convince yourself that the code is uniquely decodeable). Item (ii) is clearly prefix-free.

Lemma 7.2 *Any prefix-free code is uniquely decodeable.*

Proof Let $d_1 d_2 \ldots d_n$ be a code message comprising the codewords $d_1, d_2, \ldots, d_n$ and suppose that it is not uniquely decodeable. Suppose, further, that the ambiguity in decoding arises purely via the combination $d_{k-1} d_k d_{k+1}$, where $2 \le k \le n-1$, and that without loss of generality (check this!) both d_{k-1} and d_{k+1} are code symbols. We write $d_k = c_{k_1} c_{k_2} \ldots c_{k_m}$, then if $d_{k-1} d_k d_{k+1}$ is ambiguous, it must be that either $d_{k-1} d_k d_{k+1}$ is itself a codeword or $d_{k-1} c_{k_1} c_{k_2} \ldots c_{k_j}$ is a codeword for some

$1 \leq j \leq m$. In either case, d_{k-1} is a prefix, which contradicts our assumption that the code is prefix-free. □

Prefix-free codes do not just possess the desirable property of being uniquely decodeable, they are also instantaneously decodeable. To appreciate the implications of this consider the message 01011001 sent in the code of Example 7.6(i), which is not prefix-free. Suppose this message is being transmitted to us symbol by symbol and we try to decode each codeword as it arrives. The first symbol which arrives is a 0, which is not a codeword; the second symbol is a 1 so at this stage we have received 01. Now this might be the codeword a_2 or it could be prefix of a_4 or a_5. In fact, we have to wait until we have received the whole message (which decodes as $a_4a_3a_2$) before we can decode the first codeword unambiguously.

Prefix-free codes can always, however, be decoded instantaneously; for instance, the message 011010 in the code of Example 7.6(ii) can be decoded as $a_1a_3a_2$ as soon as each codeword arrives.

Having convinced ourselves of the desirability of prefix-free codes the next question we need to consider is whether or not it is always possible to construct these. To formalise this, suppose we have a source alphabet of n symbols, a code alphabet of r symbols, and we want to find a prefix code such that each source symbol a_i is mapped to a codeword of length l_i $(1 \leq i \leq n)$. The following result gives us a straightforward test for the existence of such a code.

Theorem 7.3 (*the Kraft–McMillan inequality*) *Given a code alphabet of r symbols, a prefix-free code with codewords of length l_i $(1 \leq i \leq n)$ exists for a source alphabet of n symbols if and only if*

$$\sum_{i=1}^{n} r^{-l_i} \leq 1. \tag{7.2}$$

Proof First suppose that (7.2) holds. We will show that it is possible to construct a prefix-free code.

Let $L = \max\{l_i; \ 1 \leq i \leq n\}$ and let n_i be the number of codewords of length i $(1 \leq i \leq L)$; then (7.2) can be written as

$$\sum_{i=1}^{L} n_i r^{-i} \leq 1.$$

Now multiply both sides of this inequality by r^L to obtain

$$\sum_{i=1}^{L} n_i r^{L-i} \leq r^L.$$

Writing the sum out explicitly and rearranging, we obtain

$$n_L \leq r^L - n_1 r^{L-1} - n_2 r^{L-2} - \cdots - n_{L-1} r$$

Since $n_L \geq 0$, the right-hand side of this inequality is non-negative. Hence, we obtain the new inequality

$$n_{L-1} r \leq r^L - n_1 r^{L-1} - n_2 r^{L-2} - \cdots - n_{L-2} r^2$$

that is

$$n_{L-1} \leq r^{L-1} - n_1 r^{L-2} - n_2 r^{L-3} - \cdots - n_{L-2} r.$$

Since $n_{L-1} \geq 0$, the right-hand side of this inequality is non-negative and so we can apply the same argument again. Continuing inductively we find that, for $0 \leq k \leq L - 1$

$$n_{L-k} \leq r^{L-k} - n_1 r^{L-k-1} - n_2 r^{L-k-2} - \cdots - n_{L-k-1} r. \qquad (*)$$

In particular, $n_1 \leq r$.

Now we construct the code. We begin with the words of length 1. There are n_1 of these and r code symbols available. Since $n_1 \leq r$, we have complete freedom in choosing these.

Now we construct the words of length 2. To ensure the code is prefix-free, we choose the first symbol from one of the $(r - n_1)$ letters remaining after the first symbol is chosen. The second symbol can then be chosen in any of r ways. By Theorem 1.1, there are then $(r - n_1)r$ possible words of length 2 available. We need to choose n_2 of these; however, by $(*)$, we have $n_2 \leq r^2 - n_1 r$ so that we are free to choose these.

We now construct the words of length 3. To ensure the prefix-free condition, we again find by Theorem 1.1 that there are $(r^2 - n_1 r - n_2)r$ available possibilities. By $(*)$, we have that $n_3 \leq r^3 - n_1 r^2 - n_2 r$ so that we are free to choose these. We now continue the argument inductively to construct our prefix-free code.

Now consider the converse, namely we must prove that if a prefix-free code can be constructed, then (7.2) must hold. We can do this by reversing the steps in the preceding argument; the details are left to Exercise 7.8.   □

Example 7.7 Determine whether or not a prefix-free code using a binary alphabet exists for a five-letter source alphabet with codewords of length 1, 2, 3, 4 and 5.

Solution We have $r = 2$ and so

$$2^{-1} + 2^{-2} + 2^{-3} + 2^{-4} + 2^{-5} = \frac{31}{32} < 1.$$

Hence by (7.2) a prefix-free code exists. It is fairly easy to construct one, for example 0, 10, 110, 1110, 11110.

7.4 Noiseless coding

Given a source alphabet of size n and a code alphabet of size r there are many possible choices of codes with different codeword sizes which satisfy the Kraft–McMillan inequality and hence yield a prefix-free code. Which of these is the best? To make this question more precise, consider the set of all possible prefix-free codes for our source alphabet S. To each code we can associate the set of lengths $\{l_1, l_2, \ldots, l_n\}$ of the codewords and we regard this set as the range of a random variable L whose law is that of S, that is, $\{p_1, p_2, \ldots, p_n\}$. We define the expectation of L in the usual way

$$\mathbb{E}(L) = \sum_{j=1}^{n} p_j l_j.$$

We say that a prefix-free code is *optimal* if it minimises $\mathbb{E}(L)$, so we are saying that the best prefix-free codes are those which have the smallest average codelength. The following theorem (which is due to Shannon) yields the, perhaps surprising, result that the minimum value of $\mathbb{E}(L)$ can be described in terms of the source entropy $H(S)$, where (recall Chapter 6)

$$H(S) = - \sum_{j=1}^{n} p_j \log(p_j).$$

Theorem 7.4 *(the noiseless coding theorem)*

(a) *For any prefix-free code we have*

$$\mathbb{E}(L) \geq \frac{H(S)}{\log(r)}$$

with equality if and only if $p_j = r^{-l_j}$ $(1 \leq j \leq n)$.

(b) *There exists a prefix-free code which satisfies*

$$\frac{H(S)}{\log(r)} \leq \mathbb{E}(L) < \frac{H(S)}{\log(r)} + 1. \tag{7.3}$$

Proof (a) We define another probability measure $\{q_1, q_2, \ldots, q_n\}$ on the source alphabet by

$$q_j = \frac{r^{-l_j}}{\sum_{i=1}^{n} r^{-l_i}}$$

then by the Gibbs inequality (Exercise 6.9) we obtain

$$H(S) \le -\sum_{j=1}^{n} p_j \log(q_j)$$

$$= -\sum_{j=1}^{n} p_j \log(r^{-l_j}) + \sum_{j=1}^{n} p_j \log\left(\sum_{i=1}^{n} r^{-l_i}\right).$$

Now by (7.2) and the fact that log is an increasing function we have

$$\log\left(\sum_{i=1}^{n} r^{-l_i}\right) \le \log(1) = 0.$$

Hence

$$H(S) \le \sum_{i=1}^{n} p_j l_j \log(r) = \mathbb{E}(L) \log(r)$$

as required.

By the equality conditions in the Gibbs inequality and the Kraft–McMillan inequality we find that we have equality if and only if $p_j = r^{-l_j}$, as required.

(b) For the right-hand inequality, choose the lengths of the codewords to be such that each $l_j (1 \le j \le n)$ is the unique whole number such that

$$r^{-l_j} \le p_j < r^{-l_j+1}.$$

It can be easily shown that the Kraft–McMillan inequality is satisfied, so that this yields a prefix-free code (see Exercise 7.10). Taking logs we obtain

$$-l_j \log(r) \le \log(p_j) < (-l_j + 1) \log(r).$$

We now multiply both sides of the right-hand part of the above inequality by p_j and sum over j to obtain

$$H(S) = -\sum_{j=1}^{n} p_j \log(p_j) > \sum_{j=1}^{n} p_j(l_j - 1) \log(r)$$

$$= (\mathbb{E}(L) - 1) \log(r),$$

from which the result follows. □

We see from Theorem 7.4(a) that if the source probabilities are of the form

$$p_j = r^{-\alpha_j} (1 \le j \le n)$$

where each α_j is a natural number, then an optimal code will be one with codeword lengths $l_j = \alpha_j (1 \le j \le n)$.

Example 7.8 Construct a binary optimal code for a source alphabet with probabilities $\frac{1}{2}$, $\frac{1}{4}$, $\frac{1}{8}$, $\frac{1}{16}$ and $\frac{1}{16}$.

Solution Since each of the probabilities is of the form 2^{-n}, where $n = 1, 2, 3, 4$
and 4, an optimal code is one with codewords of length $1, 2, 3, 4$ and 4. Specifically,
we can take

$$0, \ 10, \ 110, \ 1110, 1111.$$

Note that optimal codes are not unique in that we can interchange the role of 0 and
1 in Example 7.8 and still have an optimal code.

If the source probabilities are not of optimal form, it seems that the next best
thing we can do is to construct a prefix-free code with lengths $l_1, l_2, \ldots, l_n$ which
satisfies (7.3). To do this we see from the proof of Theorem 7.4(b) that we must
choose the l_js to be the unique natural numbers for which

$$r^{-l_j} \le p_j < r^{-l_j+1}$$

that is

$$-l_j \log(r) \le \log(p_j) < (-l_j + 1)\log(r)$$

that is

$$-\frac{\log(p_j)}{\log(r)} \le l_j < 1 - \frac{\log(p_j)}{\log(r)} \qquad (\text{or } -\log_r(p_j) \le l_j < 1 - \log_r(p_j)). \quad (7.4)$$

The coding procedure given by (7.4) is called *Shannon–Fano coding*.

Example 7.9 Find a Shannon–Fano code for a source alphabet with probabilities
0.14, 0.24, 0.33, 0.06, 0.11 and 0.12.

Solution We have by (7.4), that l_1 is the unique integer for which

$$-\log(0.14) \le l_1 < 1 - \log(0.14).$$

Hence

$$2.83 \le l_1 < 3.83 \text{ so } l_1 = 3.$$

Similar calculations yield

$$l_2 = 3, \ l_3 = 2, \ l_4 = 5, \ l_5 = 4, \ l_6 = 4.$$

Hence an appropriate prefix-free code is

$$011, \ 110, \ 00, \ 01011, \ 1111 \text{ and } 0100.$$

It is interesting to calculate the deviation of this code from the optimal. This is
given by

$$\mathbb{E}(L) - H(S) = 3.02 - 2.38 = 0.64.$$

In general, we would obviously like to make $\mathbb{E}(L)$ as close as possible to the optimum value of $\frac{H(S)}{\log(r)}$. A strong theoretical result in this direction is obtained by using the technique of *block coding*. This effectively means that instead of coding the source alphabet S, we code the *mth extension* $S^{(m)}$ of that source, where $S^{(m)}$ comprises all the source letters taken m at a time. For example:

$$\text{if} \quad S = \{A, B, C\}$$

$$S^{(2)} = \{AA, AB, BA, BB, AC, CA, BC, CB, CC\}$$

$$S^{(3)} = \{AAA, AAB, ABA, BAA, ABB, BAB, BBA, BBB, AAC,$$

$$\quad ACA, CAA, ACC, CAC, CCA, CCC, BCC, CBC, CCB,$$

$$\quad CBB, BCB, BBC, ABC, ACB, BAC, BCA, CAB, CBA\}$$

$$S^{(4)} \quad \text{will contain 81 elements.}$$

If we use binary coding, a prefix-free code for $S^{(2)}$ is

$$AA \rightarrow 0, \ AB \rightarrow 10, \ BA \rightarrow 110, \ BB \rightarrow 1110, \ AC \rightarrow 11110,$$

$$CA \rightarrow 111110, \ BC \rightarrow 1111110, \ CB \rightarrow 11111110, CC \rightarrow 11111111.$$

Let S_j be the random variable whose value is the jth symbol in the block ($1 \leq j \leq n$), so each S_j has range $\{a_1, a_2, \ldots a_n\}$. We take $S_1, S_2, \ldots, S_n$ to be i.i.d. random variables so that the source probability of the block $c_{i_1} c_{i_2} \ldots c_{i_n}$ is $p_{i_1} p_{i_2} \cdots p_{i_n}$.

Let $L^{(m)}$ be the random variable whose values are the lengths of codewords which code $S^{(m)}$. As we are coding symbols in groups of m, the quantity $\frac{\mathbb{E}(L^{(m)})}{m}$ measures the average length of codeword per source symbol. We have the following result.

Corollary 7.5 *Given any $\varepsilon > 0$, there exists a prefix-free code such that*

$$\frac{H(S)}{\log(r)} \leq \frac{\mathbb{E}(L^{(m)})}{m} < \frac{H(S)}{\log(r)} + \varepsilon.$$

Proof Given any $\varepsilon > 0$ (no matter how small), it is a well-known fact about the real numbers that there exists some $m \in \mathbb{N}$ such that $\frac{1}{m} < \varepsilon$.

Having chosen such an m, we use it for block coding as above and apply Theorem 7.3(b) to obtain a prefix-free code such that

$$\frac{H(S^{(m)})}{\log(r)} \leq \mathbb{E}(L^{(m)}) < \frac{H(S^{(m)})}{\log(r)} + 1$$

but, by Exercise 7.14, we have $H(S^{(m)}) = m H(S)$, from which the result follows.
□

Corollary 7.5 tells us that by coding in large enough blocks we can make the average length of codeword per source symbol as close to the optimal value as we like but, in practice, block coding soon becomes very complicated as the blocks increase in size. Fortunately, there is an alternative to block coding for obtaining optimal codes, called *Huffman coding*. This provides an algorithm (or recipe) for constructing optimal codes which are called *Huffman codes*. So, given a source alphabet S with probability law $\{p_1, p_2, \ldots, p_n\}$ and a code alphabet of length r, the corresponding Huffman code has average length closer to the optimal value of $\frac{H(S)}{\log(r)}$ than does any other prefix code (including those obtained via block coding with m as large as you like!). So that, in particular, Huffman codes are better than Shannon – Fano codes. For binary codes (which is all that we will consider here), Huffman's procedure is easy to carry out but difficult to describe adequately in words.

Huffman's algorithm

Write the probabilities in decreasing order – suppose, for convenience, that $p_n < p_{n-1} < \cdots < p_2 < p_1$. Merge the symbol S_n and S_{n-1} into a new symbol with probability $p_n + p_{n-1}$ so that we have a new source alphabet with $n - 1$ symbols. Now repeat the above procedure inductively until we finish with a source alphabet of just two symbols. Code these with a 0 and a 1. Now work backwards, repeating this coding procedure at each step.

Don't worry if you are completely confused. It is best to carry out Huffman coding with the aid of a 'tree diagram'. The following example should help.

Example 7.10 Find a Huffman code for the source alphabet of Example 7.8.

Solution The diagram, which should be self-explanatory, is shown in Fig. 7.5. Notice that we apply the procedure described above from left to right to reduce the code alphabet by one symbol at a time. After assigning our 0s and 1s, we then read the code backwards from right to left.

We obtain the code

$$a_1 \to 0, \ a_2 \to 10, \ a_3 \to 110, \ a_4 \to 1110, \ a_5 \to 1111.$$

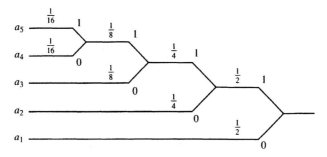

Fig. 7.5.

Note that this is precisely the code found in Example 7.8, which we know to be optimal because of the form of the input probabilities. We will not give a proof here that Huffman codes are optimal. Those who want to see one are referred to page 23 of Goldie and Pinch (see references at the end of the chapter). To see the practical advantage of Huffman coding over Shannon–Fano coding, consider the following example.

Example 7.11 Find a Huffman code for the source alphabet of Example 7.9.

Solution The tree diagram is sketched in Fig. 7.6. So we obtain the code

$$a_1 \to 01, \ a_2 \to 11, \ a_3 \to 001, \ a_4 \to 000, \ a_5 \to 101, \ a_6 \to 100.$$

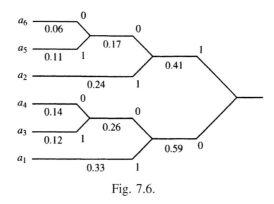

Fig. 7.6.

In this case we find

$$\mathbb{E}(L) - H(S) = 2.43 - 2.38 = 0.05$$

so the code is clearly far superior to the Shannon–Fano one found above.

We close this section by making a comment on the expression $\frac{H(S)}{\log(r)}$, which has played such an important role in this section. We define $H_r(S) = \frac{H(S)}{\log(r)}$ and call it the *r-nary entropy* so that $H_2(S) = H(S)$. Note that

$$H_r(S) = -\sum_{j=1}^{n} p_j \frac{\log(p_j)}{\log(r)} = -\sum_{j=1}^{n} p_j \log_r(p_j)$$

so $H_r(S)$ is a natural measure of entropy in contexts where every question has r possible alternative answers. We can then, via Theorem 7.4(a), interpret $H_r(S)$ as being the minimum average wordlength (in r-nary code) required to convey the information content of the source. This gives us yet another insight into the meaning of entropy.

7.5 Coding and transmission with noise – Shannon's theorem (∗)

In this section we consider the most important question in communication theory, namely by choosing an appropriate code for the source can we offset the effects of the noise and so reduce the probability of making an error to as near zero as possible, while transmitting the information at a rate arbitrarily close to the channel capacity? Surprisingly, this can be done, at least from a theoretical point of view, and below we will prove Shannon's theorem, which makes this result precise. We note from a practical viewpoint, however, that the implications of this theorem are less clear; indeed, nearly 50 years after the theorem was first presented, the required codes have still not been constructed. The reasons for this will become clearer after we have given the proof.

As the full proof of Shannon's theorem is very difficult, we are only going to present this in a special case – that of the binary symmetric channel. Even here, the level of mathematical sophistication is high and readers are invited to skip the rest of this section at a first reading.

We begin by introducing some new concepts, which we will need to formulate Shannon's theorem.

7.5.1 Decision rules

Suppose that $C = \{x_1, x_2, \ldots, x_N\}$ is the set of all possible codewords that can be transmitted across a channel, and y is the codeword received. How do we decide which of the x_js was in fact sent? We need a *decision rule*, that is a procedure whereby we can systematically assign a unique transmitted codeword to any one received. The decision rule we'll adopt here is the *maximum likelihood rule*. Consider the N conditional probabilities

$$P_y(x_j \text{ sent}) \qquad (1 \leq j \leq N)$$

where we condition on the event that y is received. We adopt the procedure that we believe that x_k was sent out if

$$P_y(x_k \text{ sent}) > P_y(x_i \text{ sent}) \qquad \text{for all } i \neq k.$$

It is shown in Exercise 7.17 that this is equivalent to the condition

$$P_{x_k \text{ sent}}(y) > P_{x_i \text{ sent}}(y) \qquad \text{for all } i \neq k.$$

If two or more codewords have equal maximum likelihoods, we decide by choosing one at random (i.e. in accordance with a uniform distribution – see Example 7.12 below).

Note that just because we decide that x_k was sent out doesn't mean that it actually was. This situation is similar to that in statistics, where accepting a hypothesis does not necessarily mean that the state of affairs which it describes is true.

The maximum likelihood procedure is not the only possible decision rule that can be used for decoding, as we will see below.

7.5.2 Hamming distance

Consider two binary codewords a and b of equal length. The *Hamming distance* $d(a, b)$ between them is simply the number of symbols in a which are different from b (or vice versa). For example, if

$$a = 00101011, \ b = 10111010,$$

then

$$d(a, b) = 3.$$

An obvious decision rule to use when transmitting binary codewords is that of minimum Hamming distance, that is the transmitted codeword is that which has minimum Hamming distance from the received codeword. We have the following result.

Lemma 7.6 *For a binary symmetric channel where $0 \leq \varepsilon < \frac{1}{2}$, the minimum Hamming distance decision rule is the same as the maximum likelihood one.*

Proof Let us suppose that a certain codeword y of length m is received. The probability that a given codeword x was sent out such that $d(x, y) = p \leq m$ is

$$P_{x \text{ sent}}(y) = \varepsilon^p (1 - \varepsilon)^{m-p}$$

$$= (1 - \varepsilon)^m \left(\frac{\varepsilon}{1-\varepsilon}\right)^p.$$

As $0 \leq \varepsilon < \frac{1}{2}$, $\frac{\varepsilon}{1-\varepsilon} < 1$, hence $P_y(x \text{ sent})$ is a maximum when p is a minimum. $\square$

7.5.3 Average error probability

Let E be the event that an error occurs, that is that our decision rule leads us to make an incorrect assumption about which symbol was sent out. We denote by $P_{x_j}(E)$, the conditional probability that an error was made given that x_j was sent out ($1 \leq j \leq N$). The *average error probability P(E)* is then given by

$$P(E) = \sum_{j=1}^{N} P_{x_j}(E) P (x_j \text{ sent out}).$$

For simplicity we will in the sequel always take a uniform distribution on the set C of codewords sent out so that

$$P(E) = \frac{1}{N} \sum_{j=1}^{N} P_{x_j}(E). \tag{7.5}$$

Clearly, the value of $P(E)$ depends upon our judgement that an error has been made, which itself depends on the particular decision rule that has been adopted.

Example 7.12 Calculate the average error probability when $\{00, 11\}$ are sent over a binary symmetric channel using a maximum likelihood decision rule.

Solution First suppose that 00 is sent out and we receive 01. As the received codeword has equal Hamming distance from both 00 and 11, we toss a coin to decide which symbol was emitted. Hence the error probability in this case is $\frac{1}{2}\varepsilon(1-\varepsilon)$. A similar argument applies to receiving 10. If 11 is received, we decode it as 11, thus obtaining a contribution ε^2 to $P_{00}(E)$. Hence, we have

$$P_{00}(E) = \varepsilon(1-\varepsilon) + \varepsilon^2 = \varepsilon.$$

An analogous argument shows that we also have $P_{11}(E) = \varepsilon$, so by 7.5, we obtain

$$P(E) = \varepsilon.$$

7.5.4 Transmission rate

For simplicity, we will assume throughout that our source emits symbols into the channel at a rate of one per second.

The *transmission rate* R is defined to be the average number of bits of information transmitted across the channel per second (the observant reader will note that R is nothing other than the mutual information between source and receiver). Clearly, we must have $0 \le R \le C$, where C is the channel capacity. Suppose that we transmit information for t seconds, then we transmit, on average, tR bits of information. If we are using a binary code, then the average total number of symbols transmitted in t seconds is $[2^{tR}]$, where for any real number x, $[x]$ is the largest integer $\le x$ (e.g. $[\pi] = 3$, $[e] = 2$, $[\sqrt{2}] = 1$).

Example 7.13 A binary symmetric channel has an error probability of 0.4. If the channel operates at capacity, what is the maximum average number of binary symbols which can be transmitted in one minute?

Solution By Example 7.2, we have

$$C = 1 - H_b(0.4) = 0.029.$$

Hence number of symbols is $[2^{0.029 \times 60}] = 3$.

We now have sufficient conceptual apparatus to express Shannon's theorem mathematically. Essentially, it tells us that we can transmit at a rate R arbitrarily close to C while keeping $P(E)$ arbitrarily small. Before diving into the nuts and bolts of the theorem we need a technical lemma.

Lemma 7.7 *Let* $0 \leq \varepsilon < \frac{1}{2}$ *and* $m \in \mathbb{N}$, *then*

$$\sum_{k=0}^{[m\varepsilon]} \binom{m}{k} \leq 2^{m H_b(\varepsilon)}. \tag{7.6}$$

Proof Using the binomial theorem we have that

$$1 = [\varepsilon + (1 - \varepsilon)]^m$$

$$= \sum_{k=0}^{m} \binom{m}{k} \varepsilon^k (1 - \varepsilon)^{m-k}$$

$$\geq \sum_{k=0}^{[m\varepsilon]} \binom{m}{k} \varepsilon^k (1 - \varepsilon)^{m-k}$$

$$= (1 - \varepsilon)^m \sum_{k=0}^{[m\varepsilon]} \binom{m}{k} \left(\frac{\varepsilon}{1 - \varepsilon}\right)^k.$$

Now, since $0 \leq \varepsilon < \frac{1}{2}$, $\left(\frac{\varepsilon}{1-\varepsilon}\right) < 1$, hence

$$\left(\frac{\varepsilon}{1 - \varepsilon}\right)^k \geq \left(\frac{\varepsilon}{1 - \varepsilon}\right)^{\varepsilon m} \quad \text{for all } 0 \leq k \leq [\varepsilon m].$$

So, we have that

$$1 \geq (1 - \varepsilon)^m \left(\frac{\varepsilon}{1 - \varepsilon}\right)^{\varepsilon m} \sum_{k=0}^{[m\varepsilon]} \binom{m}{k},$$

so

$$\sum_{k=1}^{[m\varepsilon]} \binom{m}{k} \leq [\varepsilon^{-\varepsilon}(1 - \varepsilon)^{-(1-\varepsilon)}]^m$$

$$= 2^{m H_b(\varepsilon)} \qquad\qquad \text{by Exercise 6.7.}$$

$\square$

We are now ready to prove Shannon's theorem. To simplify matters we will split the proof into two stages. First we examine the context.

We have a binary symmetric channel with error probability ε. We will impose the condition $\varepsilon < \frac{1}{2}$. This in fact involves no loss of generality as the key quantity which we deal with in the theorem is $H_b(\varepsilon)$, which is the same as $H_b(1 - \varepsilon)$. So for the case where $\varepsilon \geq \frac{1}{2}$, we can just replace it by $1 - \varepsilon < \frac{1}{2}$.

We will assume that our source S is a *Bernoulli source*, that is the random variables whose values are the symbols sent out are i.i.d.

Suppose we aim to send a codeword x over the channel which comprises d binary digits so there are 2^d possible choices of codeword available. Suppose also that for our code we only want to use M of these possibilities. Shannon had the ingenious idea of selecting these by *random coding*, that is impose a uniform distribution on the set of all possible codewords so that each is chosen with probability 2^{-d}. Hence, the probability of selecting a given M codewords to be the code is 2^{-dM}.

Now suppose that a certain codeword y is received. Let X_j be the random variable that the jth symbol in y is an error, so X_j is a Bernoulli random variable with parameter $\varepsilon (1 \leq j \leq d)$. The random variable whose values are the total number of errors in y is

$$S(d) = X_1 + X_2 + \cdots + X_d.$$

By Section 5.7, $S(d)$ is a binomial random variable with parameters d and ε and, in particular, by (5.11) we have

$$\mathbb{E}(S(d)) = \varepsilon d.$$

We now need to adopt a decision rule for decoding. The idea is to think of x – the codeword sent and y – the codeword received as vectors in a d-dimensional space, and consider within that space $\Sigma(\varepsilon d)$ – the sphere of radius εd centred on y (if you don't know about higher-dimensional geometry, just think of the case $d = 3$).

Now expand the sphere by a small amount νd, where ν is an arbitrarily small number, so we work in the sphere $\Sigma(r)$ of radius $r = d(\varepsilon + \nu)$ – the reason for this will be explained below (see Fig. 7.7).

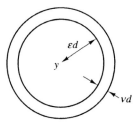

Fig. 7.7.

We adopt the following decision rule. If there is just one codeword (z, say) inside this sphere, then we decode y as z and say that z was transmitted. If there are no codewords or more than two codewords inside the sphere, we say that an error has been made.

The idea behind this decision rule is as follows. We have seen that the average Hamming distance between x and y is εd. Hence, on average, x should be somewhere within $\Sigma(\varepsilon d)$. Since the average smooths out possible outcomes in which x is outside the sphere, we expand the sphere by the amount vd in the hope of also capturing x in these cases.

Let A be the event that there are no codewords inside the sphere of radius r, and let B be the event that there are two or more. For the average error probability $P(E)$, we have

$$P(E) = P(A \cup B)$$
$$= P(A) + P(B) - P(A \cap B) \qquad \text{by (P7)}$$
$$\leq P(A) + P(B). \qquad (7.7)$$

Proposition 7.8 *Let $\delta_1 > 0$ be a fixed number (which we want to be as small as possible); then if d is chosen to be sufficiently large*

$$P(A) \leq \delta_1.$$

The proof of Proposition 7.8 depends on Chebyshev's inequality, which will be established in the next chapter. Consequently, we will postpone the proof until the end of Section 8.4.

The intuition behind the proof of Proposition 7.8 is straightforward. We have M codewords lying about somewhere in d-dimensional space (again, think of the case $d = 3$ if it helps). If there are no codewords inside our sphere, increase d so that the sphere expands. As d gets bigger, more and more of the space lies inside the sphere and it becomes less and less likely that it is empty of codewords.

Recall from Example 7.3 that the capacity C of our channel is given by $C = 1 - H_b(\varepsilon)$.

Proposition 7.9 *Let ρ and δ_2 be fixed non-negative numbers (which we want to be as small as possible) and suppose that $M = 2^{d(C-\rho)}$; then for sufficiently large d*

$$P(B) \leq \delta_2.$$

Proof Suppose that the event B is realised and there are two or more codewords in $\Sigma(r)$, where $r = d(\varepsilon + v)$. Let $x_i \in C$ be the codeword which has minimum Hamming distance from $y(1 \leq i \leq M)$. We have

$$P(B) = P\left((x_i \in \Sigma(r)) \cup \left(\bigcup(x_j \in \Sigma(r)), \ j \neq i\right)\right)$$

$$\leq P\left(\bigcup(x_j \in \Sigma(r)), \ j \neq i\right) \qquad \text{by (P6)}$$

$$\leq \sum_{j=1, j\neq i}^{M} P(x_j \in \Sigma(r))$$

$$= (M-1)P(x_j \in \Sigma(r)) \text{ for some } 1 \leq j \leq M. \qquad \text{(i)}$$

as the codewords are chosen randomly.

Now $x_j \in \Sigma(r)$ if x_j has up to $[r]$ errors so, for each $1 \leq k \leq M$, as the probability of k errors is $\frac{1}{2^d}\binom{d}{k}$, we have

$$P(x_j \in \Sigma(r)) = \frac{1}{2^d} \sum_{k=0}^{[r]} \binom{d}{k}$$

$$\leq \frac{1}{2^d} 2^{d H_b(\varepsilon + \nu)} = 2^{-d(1 - H_b(\varepsilon + \nu))}. \qquad \text{(ii)} \qquad \text{by (7.6)}$$

So, on combining (i) and (ii), we obtain

$$P(B) \leq (M-1)2^{-d(1 - H_b(\varepsilon + \nu))}$$

$$\leq M 2^{-d(1 - H_b(\varepsilon + \nu))}.$$

Now we have

$$M = 2^{d(C - \rho)} = 2^{d(1 - H_b(\varepsilon) - \rho)}.$$

Hence

$$P(B) \leq 2^{d(H_b(\varepsilon + \nu) - H_b(\varepsilon) - \rho)}.$$

Now recall the graph of H_b which we sketched in Section 6.2 and note that for $\varepsilon < \frac{1}{2}$ this is an increasing (and continuous) function; hence we can, by choosing ν sufficiently small, ensure that $H_b(\varepsilon + \nu) - H_b(\varepsilon) < \rho$ so that $b > 0$, where

$$b = \rho - (H_b(\varepsilon + \nu) - H_b(\varepsilon)).$$

Now if we take d to be sufficiently large, no matter how small δ_2 is, we can make $2^{-db} < \delta_2$ and we have

$$P(B) \leq \delta_2, \text{ as required.}$$

$$\square$$

Theorem 7.10 (*Shannon's fundamental theorem*) *Given* $\delta, \rho > 0$ *(which we can take as small as we like), we can find a code such that if the transmission rate over* *a binary symmetric channel* $R = C - \rho$, *then*

$$P(E) < \delta.$$

Proof Just take d to be large enough for the hypotheses of both Propositions 7.8 and 7.9 to be realised and put $\delta = \delta_1 + \delta_2$; then by (7.7)

$$P(E) \leq \delta_1 + \delta_2 = \delta.$$

$\square$

Notes:

(i) Although we have only stated and proved Shannon's theorem for a binary symmetric channel it does in fact hold for any channel. The proof is essentially along the same lines as above but is somewhat harder.

(ii) It is possible to improve Shannon's theorem by showing that, instead of the average error probability, we can make the maximum error probability as small as we like where this latter probability is defined to be

$$\max_{1 \leq j \leq M} P_{x_j}(E)$$

(iii) Shannon's theorem has a converse, which we will not prove here, which states that it is not possible to make the error probability arbitrarily small if we transmit at a rate above the channel capacity.

The following quotation by D. Slepian, which first appeared in 1959, still gives an appropriate description of the status of Shannon's theorem, it can be found on p.166 of Reza's book (see references):

From the practical point of view, the fundamental theorem contains the golden fruit of the theory. It promises us communication in the presence of noise of a sort that was never dreamed possible before: perfect transmission at a reasonable rate despite random perturbations completely outside our control. It is somewhat disheartening to realise that today, ten years after the first statement of this theorem, its content remains only a promise – that we still do not know in detail how to achieve these results for even the most simple non-trivial channel.

In essence, Shannon's result is of a type known in mathematics as an 'existence theorem'. It tells us that somewhere ('over the rainbow'?) there is a wonderful code with miraculous properties. The downside is that the proof gives us no indication as to how this code might be constructed in practice.

7.6 Brief remarks about the history of information theory

Information theory is a very young subject – in fact a child of the twentieth century. The idea of using logarithms to measure information content dates back to the work of Hartley in 1928. However, the most important works establishing the subject were the seminal papers of Claude Shannon beginning with his 'A mathematical theory of communication' published in 1948. Most of the ideas treated in this chapter and many of those in the preceding one – the use of entropy, channel capacity, the noiseless coding theorem and of course the fundamental theorem – all have their origins in this paper. Indeed, much of the subsequent development of the theory has been devoted to the reformulation and refinement of Shannon's original ideas. The reader might be interested to note that, at the time he did his original work, Shannon was a communications engineer working for the Bell Telephone Laboratories and not a mathematics professor in a university!

After Shannon's paper was published, a great deal of work was carried out on sharpening the proof of his fundamental theorem – in particular by Fano, Feinstein, Shannon himself and Wolfowitz. It is always difficult to try to sum up more recent developments but of these perhaps the most exciting are the application of information theory to statistical inference and recent developments in coding and cryptography (see the books by Blahut and Welsh, referenced below, respectively).

Exercises

7.1. Recall the binary erasure channel of Exercise 4.16. Obtain expressions for all the mutual informations $I(0, 0)$, $I(0, E)$ and $I(1, 1)$. Find also the mutual information $I(R, S)$ between source and receiver and hence deduce the channel capacity C.

7.2. Using the results of Chapter 4, confirm that the receiver's entropy $H(R)$ for a binary symmetric channel is $H_b(x)$, where $x = \varepsilon + p - 2\varepsilon p$. Show that $x = \frac{1}{2}$ if and only if $p = \frac{1}{2}$, provided that $\varepsilon \neq \frac{1}{2}$. Why does this latter restriction have no bearing on the calculation of channel capacity in Section 7.2? (See Exercise 4.15.)

7.3. Consider again the binary errors and erasure channel of Exercise 4.17. Obtain:

(a) $H(R)$,
(b) $H_S(R)$,
(c) $I(S, R)$.

7.4. Use differentiation to find the maximum value of

$$f(y) = -y \ln(y) - (1 - y - \rho) \ln(1 - y - \rho)$$

where ρ is constant. Hence, find the channel capacity of the binary errors and erasures channel and show that it is attained when $p = \frac{1}{2}$.

7.5. The channel of Exercise 4.22 is an example of a 'cascade'. Calculate $I(A, B)$ and $I(A, C)$ and confirm that $I(A, B) - I(A, C) > 0$.

7.6. In general, suppose a source with alphabet $A = \{a_1, a_2, \ldots, a_n\}$ emits symbols into a channel to a receiver with alphabet $B = \{b_1, b_2, \ldots, b_m\}$, which then itself acts as a source for transmitting symbols across a second channel to a receiver with alphabet $C = \{c_1, c_2, \ldots, c_r\}$. This set-up is called a *cascade* if

$$P_{a_i \cap b_j}(C_k) = P_{b_j}(C_k)$$

for all $1 \le i \le n,\ 1 \le j \le m,\ 1 \le k \le r$.

(i) Show that each

$$P_{b_j \cap c_k}(a_i) = P_{b_j}(a_i).$$

(ii) Use the result of (i) and Gibb's inequality to deduce that

$$H_C(A) \le H_B(A).$$

(iii) Hence, deduce that

$$I(A, B) \ge I(A, C).$$

The result of (iii) is sometimes called the *data processing theorem* – repeated processing of information cannot increase the amount transmitted.

7.7. (for those interested in biology) The genetic code for the DNA of the bacterium *Micrococcus Lysodeiktus* has source probabilities

$$P(C) = P(G) = 0.355,\ \ P(A) = P(T) = 0.145$$

while for *E. Coli*

$$P(A) = P(G) = P(C) = P(T) = 0.25.$$

Which organism would you expect to manifest greater complexity?

7.8. Complete the proof of Theorem 7.3, that is show that if a prefix-free code exists using a code alphabet of r symbols with codeword lengths $l_1, l_2, \ldots, l_n$ for an alphabet of n symbols, then the Kraft–McMillan inequality is satisfied.

7.9. The trinary code consists of the symbols $\{0,1,2\}$. It is required to construct a trinary code with words of length 1, 2, 2, 2, 2, 3, 3, 3, 3, 4, 4. Show that it is possible to construct a prefix-free code and find such a code.

7.10. Show that Shannon–Fano coding yields a prefix-free code.

7.11. A source alphabet of seven symbols has probabilities $\frac{1}{3}, \frac{1}{3}, \frac{1}{9}, \frac{1}{9}, \frac{1}{27}, \frac{1}{27}, \frac{1}{27}$. Find an optimal code using a trinary alphabet.

7.12. A source with probabilities 0.27, 0.31, 0.14, 0.18 and 0.10 is to be coded into binary; find:

(a) a prefix-free code using Shannon–Fano coding,
(b) an optimal code using Huffman coding.

7.13. The efficiency of a code, η, is defined by $\eta = \frac{H_r(S)}{\mathbb{E}(L)}$. Find the efficiency of all three codes in (11) and (12) above.

7.14. Show that if we take the mth extension $S^{(m)}$ of a source, then

$$H(S^{(m)}) = mH(S).$$

7.15. (a) For a source with probabilities 0.02, 0.09, 0.15, 0.21, 0.26 and 0.27, show that it is possible to carry out the Huffman procedure in two different ways. Write down the code in each case. Is either of the codes more optimal than the other?

(b) The *redundancy* ξ of a code is defined by $\xi = 1 - \eta$, where η is the efficiency (see Exercise 7.13 above). Calculate the redundancy of the optimal code in (a) above.

7.16. One way of detecting transmission errors is to use a *parity check*. A simple example of this is to add an extra digit on the end of each codeword to ensure that the total number of 1s is even, for example $001 \rightarrow 0011$, $000 \rightarrow 0000$. Construct such an *error-detecting code* for the optimal code found in Exercise 7.15 and calculate its redundancy.

7.17. Let ξ_1 be the redundancy of a code with average length $\mathbb{E}(L_1)$ and let ξ_2 be the redundancy of this code when it is extended to include a parity check as in Exercise 7.16 so that its average length changes to $\mathbb{E}(L_2)$. Show that

$$\xi_2 = \xi_1 + \frac{H_r(S)}{\mathbb{E}(L_1)\mathbb{E}(L_2)},$$

and hence deduce that $\xi_2 \geq \xi_1$.

7.18. Convince yourself that the Hamming distance satisfies the triangle inequality, that is, if x, y and z are three binary codewords of length n, then

$$d(x, z) \leq d(x, y) + d(y, z).$$

Further reading

Most of the material in this chapter is standard and can be found in most, if not all, books on information theory. Those that I found most helpful were N. Abramson *Information Theory and Coding* (McGraw Hill, 1963), R. Ash *Information Theory* (Dover, 1965), R. Hamming *Coding and Information Theory* (Prentice-Hall, 1980) and F. Reza *An Introduction to Information Theory* (McGraw-Hill, 1961). The books by Shannon and Weaver and Rényi mentioned at the end of the last chapter remain relevant to this one, as does Khintchin, although it is much harder.

Among more recent texts, D. Welsh, *Codes and Cryptography* (Clarendon Press, Oxford, 1988) contains a very nice succinct proof of Shannon's fundamental theorem, as well as an excellent introduction to modern coding theory. For those who want a more advanced (but not unreadable) approach to information theory, try *Communication Theory* by C. Goldie and R. Pinch (Cambridge University Press, 1991), if they are theoretically minded, or *Principles and Practice of Information Theory* by R. Blahut (Addison-Wesley, 1987), for a more applied context.

A classic reference for the scope of information theory within science is L. Brillouin *Science and Information Theory* (Academic Press, 1962). A fascinating description of an information theoretic approach to genetics is *Information Theory and the Living System* by L. Gatlin (Columbia University Press, 1972). The information theoretic distance discussed in Exercise 6.14 plays a fundamental role in Gatlin's theory. An interesting critique of the application of information theory to the behaviour of living systems can be found in Appendix 2 ('What is Information?') of *The Way* by E. Goldsmith (Rider, 1992).

8

Random variables with probability density functions

8.1 Random variables with continuous ranges

Since Chapter 5, we have been concerned only with discrete random variables and their applications, that is random variables taking values in sets where the number of elements is either finite or ∞. In this chapter, we will extend the concept of random variables to the 'continuous' case wherein values are taken in $\mathbb{R}$ or some interval of $\mathbb{R}$.

Historically, much of the motivation for the development of ideas about such random variables came from the theory of errors in making measurements. For example, suppose that you want to measure your height. One approach would be to take a long ruler or tape measure and make the measurement directly. Suppose that we get a reading of 5.7 feet. If we are honest, we might argue that this result is unlikely to be very precise – tape measures are notoriously inaccurate and it is very difficult to stand completely still when you are being measured.

To allow for the uncertainty as to our true height we introduce a random variable X to represent our height, and indicate our hesitancy in trusting the tape measure by assigning a number close to 1 to the probability $P(X \in (5.6, 5.8))$, that is we say that our height is between 5.6 feet and 5.8 feet with very high probability. Of course, by using better measuring instruments, we may be able to assign high probabilities for X lying in smaller and smaller intervals, for example $(5.645, 5.665)$; however, since the precise location of any real number requires us to know an infinite decimal expansion, it seems that we cannot assign probabilities of the form $P(X = 5.67)$. Indeed, there is no measuring instrument that can distinguish, for example, between the two numbers 5.67 and $5.67 + 10^{-47}$, so how would we know which is the correct height? We now begin to formalise these ideas by fixing an interval (a, b) of the real line $((a, b)$ could be the whole of $\mathbb{R}$ if we take $a = -\infty$ and $b = \infty)$. We introduce a probability space $(S, \mathcal{B}(S), P)$ and define a *random variable with range (a,b)* to be a mapping X from S to (a, b). In order to be able to describe probabilities of

X taking values in certain regions, we introduce the Boolean algebra $\mathcal{I}(a, b)$ of all broken lines within (a, b). For each $G \in \mathcal{I}(a, b)$ we define

$$p_X(G) = P(X \in G). \qquad (8.1)$$

Notes

(i) Some readers may like to bear in mind that $P(X \in G)$ is shorthand for $P(\{\omega \in S;\ X(\omega) \in G\})$. For such a probability to exist, it is essential that the set $X^{-1}(G) \in \mathcal{B}(S)$, where $X^{-1}(G) = \{\omega \in S;\ X(\omega) \in G\}$. In general, there is no reason why this should be the case and, consequently, advanced books on probability theory always restrict the definition of 'random variable' to those mappings which have this property. In the more general context of measure theory, these are called 'measurable functions'.

(ii) Random variables with range (a, b) are sometimes called 'continuous' random variables. It is important to appreciate that this is not the same thing as a 'continuous function' taking values in (a, b).

Generalising the argument of Lemma 5.1, we have the following Lemma.

Lemma 8.1 p_X *is a probability measure on* $\mathcal{I}(a, b)$.

Proof Let G_1 and G_2 be disjoint sets in $\mathcal{I}(a, b)$; then

$$
\begin{aligned}
p_X(G_1 \cup G_2) &= P(X \in G_1 \cup G_2) \\
&= P((X \in G_1) \cup (X \in G_2)) \\
&= P(X \in G_1) + P(X \in G_2) \qquad \text{since } P \text{ is a measure} \\
&= p_X(G_1) + p_X(G_2)
\end{aligned}
$$

so p_X is a measure. To see that p_X is in fact a probability measure, note that

$$p_X((a, b)) = p(X \in (a, b)) = 1.$$

$\square$

We call p_x the *probability law* of the random variable X. The *cumulative distribution* also makes sense for these random variables and is defined by

$$
\begin{aligned}
F(x) &= p_X((a, x)) \quad \text{for } a \le x \le b \\
&= P(X \le x).
\end{aligned}
$$

We leave to Exercise 8.2 the appropriate generalisation of Lemmas 5.2 and 5.3 in this context (see also Equation (8.1) below). In the discussion of height measurements at the beginning of this chapter, we argued that it made no sense, in practice, to speak of a random variable with a continuous range taking a precise value. In general, we say that a random variable X is *distributed continuously* if $p_X(\{x\}) = 0$ for all

$x \in (a, b)$. All the examples of random variables we will deal with in this chapter will be distributed continuously. It turns out that the cumulative distribution of such random variables is a very useful tool in their analysis, as the following result shows.

Lemma 8.2 *If X is a continuously distributed random variable on (a,b), then its law p_X is completely determined by its cumulative distribution.*

Proof Let $J \in \mathcal{I}(a, b)$. Since X is distributed continuously, we can take $J = I_1 \cup I_2 \cup \cdots \cup I_n$ for some $n \in \mathbb{N}$, where each $I_k = (x_k, y_k)$ with $a \le x_1 < y_1 < x_2 < y_2 < \cdots < x_n < y_n \le b$.

By (P3) (p. 37) and Lemma 8.1, we have

$$p_X(J) = p_X(I_1) + p_X(I_2) + \cdots + p_X(I_n).$$

Now write each $I_k = (a, y_k) - (a, x_k)$ and use (P4) (p. 37) and the definition of F to obtain

$$p_X(J) = F(y_1) - F(x_1) + F(y_2) - F(x_2) + \cdots + F(y_n) - F(x_n).$$

$\square$

Inspired by Lemma 8.2, we often find it convenient to characterise the distributions of continuously distributed random variables by means of F rather than p_X. In fact, the formula obtained in the last line of the proof is very useful in practice when J is itself just an open interval (x, y), for, if we recall that $p_X(x, y) = P(X \in (x, y)) = P(x < X < y)$, then we have

$$P(x < X < y) = F(y) - F(x). \tag{8.2}$$

Notes

(i) We should bear in mind that the definition of random variables taking values in an interval is broad enough to include all the discrete random variables discussed in Chapter 5 but, of course, none of these is continuously distributed.

(ii) Any random variable with range $(a, b) \subset \mathbb{R}$ can always be extended to a random variable with range the whole of R by considering p_X as a probability measure on $\mathcal{I}(\mathbb{R})$ with

$$p_X(\overline{(a, b)}) = 0.$$

8.2 Probability density functions

Let X be a continuously distributed random variable taking values in (a, b) with cumulative distribution F. Almost all the important examples of random variables

which we meet in elementary probability are of the following type – there exists a
function f defined on (a, b) such that $f(x) \geq 0$ for all $a \leq x \leq b$ and

$$F(x) = \int_a^x f(y)dy; \tag{8.3}$$

f is called the *probability density function (pdf)* of the random variable X. By
(P2) (p.37), $P(X \in (a, b)) = p_X((a, b)) = 1$; hence, for a function f (for which
$f(x) \geq 0$ for all $a \leq x \leq b$) to be a pdf of a random variable X with range (a, b)
we must have

$$\int_a^b f(y)dy = 1. \qquad \text{(Compare this with Equation (5.1).)}$$

It follows by the fundamental theorem of calculus applied to (8.3) that if X has a
pdf f, then its cumulative distribution F is differentiable with derivative

$$F'(x) = f(x) \tag{8.4}$$

and since the derivative of F is unique it follows that a random variable cannot have more than one pdf. By Lemma 8.2 we see that if a pdf is given, then
it completely determines the law p_X. Intuitively, a pdf establishes a relationship
between 'probability' and 'area under a curve', as Fig. 8.1 shows. We will investigate this idea in greater detail in the next section. Notice that if $a \leq x < y \leq b$,
then

$$P(x < X < y) = \int_x^y f(t)dt. \tag{8.5}$$

Example 8.1 [the uniform random variable] This is the continuous analogue of the
discrete uniform distribution discussed in Section 5.2. Let $a, b \in \mathbb{R}$ with $b \geq a$;
we say that X has a *uniform distribution* on (a, b) if (see Fig. 8.2)

$$f(x) = \frac{1}{b - a} \qquad \text{for all } a \leq x \leq b.$$

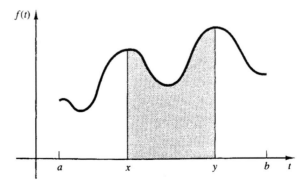

Fig. 8.1.

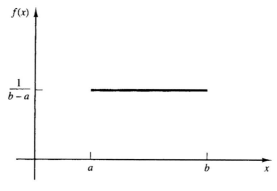

Fig. 8.2.

If X is such a random variable, we write $X \sim U(a, b)$.

To compute the cumulative distribution, we use (8.3) to obtain

$$F(x) = \int_a^x \frac{1}{b-a} dt = \frac{x-a}{b-a}.$$

Example 8.2 A uniformly distributed random variable takes values in the interval $(1, 9)$. What is the probability that its value is greater than 2 but does not exceed 6?

Solution Using the formula for F obtained above we find that

$$F(x) = \frac{1}{8}(x - 1) \text{ when } X \sim U(1, 9).$$

Now by (8.1) we find

$$P(2 \leq X \leq 6) = F(6) - F(2) = \frac{1}{8}(6 - 1) - \frac{1}{8}(2 - 1)$$
$$= \frac{1}{2}.$$

Example 8.3 [the exponential distribution] Given $\lambda > 0$, we say that X has an *exponential distribution* with parameter λ if it has range $[0, \infty)$ and pdf

$$f(x) = \lambda e^{-\lambda x}.$$

If X is such a random variable, we write $X \sim \mathcal{E}(\lambda)$. The graph of the pdf is shown in Fig. 8.3.

Using (8.1) we find that

$$F(x) = \int_0^x \lambda e^{-\lambda t} dt = 1 - e^{-\lambda x}.$$

The exponential distribution is frequently applied to describe the time to failure of electronic devices. As we will see in Example 8.5, λ^{-1} can then be interpreted as the average failure rate.

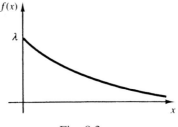

Fig. 8.3.

Example 8.4 Given that $\lambda = 1$, what is the probability that a solid state device lasts longer than two hours?

Solution $P(X > 2) = 1 - F(2) = e^{-2} = 0.135.$

If X and Y are two random variables with continuous ranges, α is a scalar and g is a real-valued function, we can form the new random variables $X + Y$, αX, XY and $g(X)$, just as we did in Chapter 5 in the discrete case.

If X has pdf f, we define its *expectation* $\mathbb{E}(X)$ by

$$\mathbb{E}(X) = \int_a^b x f(x) \mathrm{d}x \tag{8.6}$$

and, just as in the discrete case, we often write $\mu = \mathbb{E}(X)$.

More generally, we have

$$\mathbb{E}(g(X)) = \int_a^b g(x) f(x) \mathrm{d}x. \tag{8.7}$$

In particular, the *nth moment* of X is

$$\mathbb{E}(X^n) = \int_a^b x^n f(x) \mathrm{d}x \tag{8.8}$$

and the variance Var(X) is defined by

$$\mathrm{Var}(X) = \mathbb{E}((X - \mu)^2)$$

$$= \int_a^b (x - \mu)^2 f(x) \mathrm{d}x \tag{8.9}$$

and we sometimes write $\sigma^2 = \mathrm{Var}(X)$ and call σ the standard deviation of X.

Example 8.5 Find the expectation of $X \sim \mathcal{E}(\lambda)$.

Solution Using (8.6) and the pdf of Example 8.3 above we find that

$$\mathbb{E}(X) = \int_0^\infty x \cdot \lambda e^{-\lambda x} dx$$

$$= [-xe^{-\lambda x}]_0^\infty + \int_0^\infty e^{-\lambda x} dx \qquad \text{using integration by parts}$$

$$= \frac{1}{\lambda}.$$

Notes

(i) We always assume that our random variables are such that each of the expressions in (8.6)–(8.9) are finite. This is not, in fact, always the case – see Exercise 8.10 for an example.

(ii) If we compare (8.6)–(8.9) with the analogous expressions in Chapter 5, we see that they are of a similar form, the main differences being that the sum has been replaced by an integral, the value of the discrete variable by that of the continuous one and the probability p_j by '$f(x)dx$'. An attempt to explain the logic of these changes is made in the next section.

8.3 Discretisation and integration (∗)

In this section, we will try to fulfil the promise of note (ii) above and explain why Formula (8.7) is a 'natural' extension of the discrete version in Chapter 5.

An intuitive way of imagining how random variables with continuous ranges come about is to think of a (sequence of) random variable(s) taking more and more values so that these 'merge' into a continuous line. If we draw the probability histogram, we can then begin to see the pdf emerge as a curve which interpolates between the heights of each of the blocks. As an important motivating example, Figs. 8.4

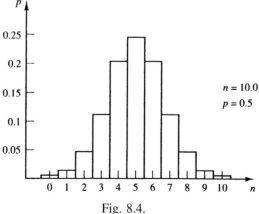

Fig. 8.4.

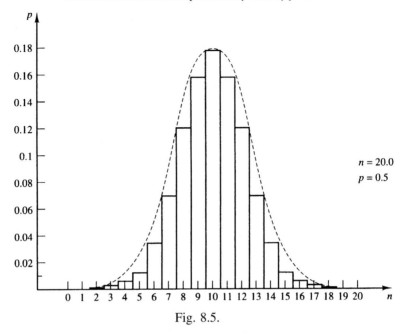

Fig. 8.5.

and 8.5 show probability histograms for the binomial distributions with $p = 0.5$ and $n = 10$, $n = 20$ respectively. In this case, the pdf which emerges is that of the normal distribution, which we will examine in some detail in Section 8.5 below.

Using the above discussion as motivation, we now concentrate on the business of justifying (8.7). We begin by remembering how (definite) integration works. Suppose that we want to calculate the area under the curve $y = h(x)$ between $x = a$ and $x = b$, as is shown in Fig. 8.6.

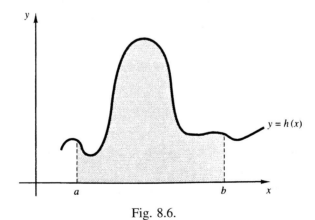

Fig. 8.6.

We first construct an approximation to the required area as follows. We form a partition $\mathcal{P}$ of the interval $[a, b]$ by dividing it into n (not necessarily equal) intervals of the form $[x_{n-1}, x_n]$, where $x_0 = a$ and $x_n = b$. Choose points ξ_j, where, for $1 \leq j \leq n$, each $\xi_j \in [x_{j-1}, x_j]$ and define

$$S(\mathcal{P}) = \sum_{j=1}^{n} h(\xi_j) \Delta x_j,$$

where $\Delta x_j = x_j - x_{j-1}$. $S(\mathcal{P})$ is then the area given in Fig. 8.7.

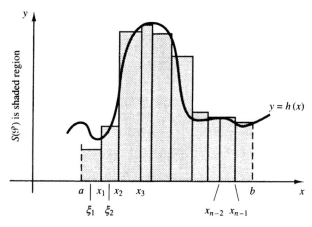

Fig. 8.7.

The idea now is to increase the number of terms n in the partition while letting each Δx_j get smaller and smaller. As this happens, the quantity $S(\mathcal{P})$ gets closer and closer to the required area. From a mathematical point of view this involves taking a limit, but as the details are quite complicated we will just write the result as

$$\int_a^b h(x) \mathrm{d}x = \lim S(\mathcal{P}),$$

and leave the reader to consult a textbook on analysis for the precise meaning of the limit.

Now we want to apply these ideas to random variables. Let X be a random variable with range $[a, b]$ and pdf f. Let $\mathcal{P}$ be a partition of $[a, b]$ as above. We construct a discrete random variable $\hat{X}$ as follows: the range of $\hat{X}$ is the n values $\xi_1, \xi_2, \ldots, \xi_n$ defined above and the probability law of $\hat{X}$ is $\{p_1, p_2, \ldots, p_n\}$, where

$$p_j = \int_{x_{j-1}}^{x_j} f(x) \mathrm{d}x = F(x_j) - F(x_{j-1}).$$

The random variable $\hat{X}$ is called a *discretisation* of X.

Now the expectation of $g(\hat{X})$ is given by

$$\mathbb{E}(g(\hat{X})) = \sum_{j=1}^{n} g(\xi_j) p_j$$

$$= \sum_{j=1}^{n} g(\xi_j)(F(x_j) - F(x_{j-1}))$$

$$= \sum_{j=1}^{n} g(\xi_j) \frac{F(x_j) - F(X_{j-1})}{\Delta x_j} \Delta x_j. \qquad (\#)$$

Now since, by the definition of the derivative and (8.4) we have

$$\lim_{\Delta x_j \to 0} \frac{F(x_j) - F(x_{j-1})}{\Delta x_j} = F'(x_j) = f(x_j);$$

we find that when we take limits in (#), we obtain

$$\lim \mathbb{E}(g(\hat{X})) = \int_a^b g(x) f(x) \mathrm{d}x = \mathbb{E}(g(X))$$

as was required.

8.4 Laws of large numbers

In this section, we aim to return to the study of sums of i.i.d. random variables which we began in Chapter 5, and also complete the discussion of the relationship between probability and relative frequency which began in Chapter 4.

First, we need to extend the ideas of i.i.d. random variables to the context of this chapter. Let X and Y be random variables with ranges (a, b) and (c, d) respectively. Clearly, the definition given in (5.16) is not appropriate in this context but the equivalent condition obtained in Exercise 5.16 gives us a clue as to how to proceed. We say that X and Y are *(probabilistically) independent* if for all $A \in \mathcal{I}(a, b)$ and all $B \in \mathcal{I}(c, d)$, we have

$$P((X \in A) \cap (Y \in B)) = P(X \in A)P(Y \in B). \qquad (8.10)$$

X and Y are said to be *identically distributed* if they have the same range (a, b) and

$$P(X \in A) = P(Y \in A) \quad \text{for all } A \in \mathcal{I}(a, b).$$

Just as in the discrete case, we can now consider i.i.d random variables with continuous ranges.

Let $X_1, X_2, \ldots$ be a family of i.i.d. random variables with common mean μ and common variance σ^2. We form their partial sums,

$$S(n) = X_1 + X_2 + \cdots + X_n,$$

and, just as in (5.8), we find that $\mathbb{E}(S(n)) = n\mu$ and $\text{Var}(S(n)) = n\sigma^2$. We now wish to investigate the 'asymptotic behaviour' of the random variables $\overline{X}(n) = \frac{S(n)}{n}$ as n gets larger and larger. To motivate this analysis, it may be useful to re-read the section on elementary statistical inference at the end of Section 5.6.

Intuitively, the values of the random variable $\overline{X}(n)$ are all possible sample means based on samples of size n. As n gets larger and larger, we should be getting more and more information in our samples about the underlying population. Hence, the values of $\overline{X}(n)$ with increasing n should be getting closer and closer to the population mean μ. So as n gets larger we would expect $|\overline{X}(n) - \mu|$ to get smaller. The next two results give a more precise meaning to these ideas.

Theorem 8.3 *(Chebyshev's inequality) Let X be a random variable with range $\mathbb{R}$ and let $c \geq 0$ be arbitrary; then*

$$P(|X - \mu| \geq c) \leq \frac{\sigma^2}{c^2}. \tag{8.11}$$

Proof Let $A = \{x \in R; |x - \mu| \geq c\}$. By (8.9) and using the fact that $\mathbb{R} = A \cup \overline{A}$, we have

$$\sigma^2 = \int_{-\infty}^{\infty} (x - \mu)^2 f(x) dx$$

$$= \int_A (x - \mu)^2 f(x) dx + \int_{\overline{A}} (x - \mu)^2 f(x) dx$$

$$\geq \int_A (x - \mu)^2 f(x) dx$$

$$\geq c^2 \int_A f(x) dx$$

$$= c^2 p_X(A)$$

$$= c^2 P(|X - \mu| \geq c) \qquad \text{by (8.1)}$$

and the result follows. □

Notes

(i) Recalling that probability is a measure, we should note that the left-hand side of (8.11) is short for

$$P(\{\omega \in S; |X(\omega) - \mu| \geq c|\}).$$

(ii) You should convince yourself that $A = (-\infty, \mu - c] \cup [\mu + c, \infty)$ so that the integral over A in the above proof is a sum of integrals over these two intervals (see Exercise 8.15).

(iii) In the proof of Theorem 8.3, we assumed that X has a density f – this is not necessary in general. In particular, the proof carries over easily to the discrete case (Exercise 8.16). We should bear this in mind for the next result.

(iv) Even if X is not distributed continuously, we see that (8.11) implies

$$P(|X - \mu| > c) \leq \frac{\sigma^2}{c^2},$$

as $\{\omega \in S; |X(\omega) - \mu| > c|\} \subseteq \{\omega \in S; |X(\omega) - \mu| \geq c|\}$.

Corollary 8.4 *(the weak law of large numbers) Let $X_1, X_2, \ldots$ be a sequence of i.i.d. random variables with common mean μ and variance σ^2. Consider the random variables defined above as $\overline{X}(n) = \frac{S(n)}{n}$; then for all $\varepsilon > 0$ we have*

$$P(|\overline{X}(n) - \mu| > \varepsilon) \leq \frac{\sigma^2}{\varepsilon^2 n}. \tag{8.12}$$

Proof Apply Chebyshev's inequality to the random variable $\overline{X}(n)$ and use $\mathbb{E}(\overline{X}(n)) = \mu$ and $\mathrm{Var}(\overline{X}(n)) = \frac{\sigma^2}{n}$. □

By (8.12), we deduce that

$$\lim_{n \to \infty} P(|\overline{X}(n) - \mu| > \varepsilon) = 0.$$

(Bear in mind note (i) above when interpreting this statement.)

The weak law tells us that as n gets larger and larger it becomes increasingly unlikely that the values of $\overline{X}(n)$ will differ appreciably from μ.

We now apply the weak law to the analysis of relative frequencies. Suppose we are carrying out a repeatable experiment whose results can be described as 'successes' or 'failures'. Let X_j be the Bernoulli random variable defined by

$$X_j = 1 \qquad \text{if the } j\text{th experiment is a success,}$$
$$= 0 \qquad \text{if it is a failure.}$$

We denote $P(X_j = 1) = p$. Remember that in practice p is unknown and so we try to estimate it by repeating the experiment a large number of times and calculating the relative frequency. We define the random variable $\overline{X}(n) = \frac{S(n)}{n}$ as usual and recognise that the values of $\overline{X}(n)$ are precisely the relative frequency of successes in n experiments, which we denoted by f_n in Chapter 4. For example, if we repeat the experiment five times and obtain three successes, we will have $\overline{X}(n) = \frac{3}{5}$.

Recalling that for the Bernoulli random variables X_j we have $\mu = p$, we find that the weak law tells us that for every $\varepsilon > 0$, we have

$$\lim_{n \to \infty} P(|f_n - p| > \varepsilon) = 0. \tag{8.13}$$

Many mathematicians and philosophers have tried to use this result to define the probability of success p as the limit of relative frequencies; however, we need to look at (8.13) more carefully. What it is telling us is that as n increases, the *probability* of the event in which the difference between f_n and p is appreciable is becoming smaller and smaller. The crux of the matter is the word *probability* above – in order to use the law of large numbers to define probability as the limit of relative frequencies, we need to already know what *probability* is. Consequently the argument is circular and hence invalid. Equation (8.13) does, however, give a mathematical legitimacy to the use of relative frequencies to *approximate* probabilities and this is clearly of great practical value in itself.

There is a more powerful result than Theorem 8.3, which is called the *strong law of large numbers*. Its proof goes beyond this book but we can at least state it; it says that under the same hypotheses as were required for Theorem 8.3, we have

$$P\left(\lim_{n \to \infty} \overline{X}(n) = \mu \right) = 1.$$

This statement is stronger in the sense that the weak law, Theorem 8.3, can be derived as a consequence of it.

Finally, we complete this section by returning to the proof of Shannon's fundamental theorem given in Section 7.5 and providing the missing proof of Proposition 7.8. We recommend readers to re-familiarise themselves with the context and notation.

Proof of Proposition 7.8 We begin by noting that the event A is precisely the event $(S(d) > r)$, where $S(d)$ is a binomial random variable with parameters ε and d and where $r = d(\varepsilon + v)$. So

$$P(A) = P(S(d) > r) = P(S(d) - \varepsilon d > vd)$$
$$\leq P(|S(d) - \varepsilon d| > vd).$$

Now apply Chebyshev's inequality (see Exercise 8.16) with $X = S(d)$, $\mu = \varepsilon d$, $\sigma^2 = \varepsilon(1 - \varepsilon)d$ and $c = vd$ to obtain

$$P(A) \leq \frac{\varepsilon(1 - \varepsilon)}{vd}$$

and this can be made smaller than any given δ_1 by taking d large enough. □

8.5 Normal random variables

In this section, we will begin the study of what is perhaps the most important class of random variables in probability theory. These are called *normal* or *Gaussian* (after the great mathematician Carl Friedrich Gauss (1777–1855)) and there is a different random variable for each value of the two parameters $\mu \in \mathbb{R}$ and $\sigma > 0$. The pdf for such a random variable is given by the formula

$$f(x) = \frac{1}{\sigma \sqrt{(2\pi)}} \exp\left\{-\frac{1}{2}\left(\frac{x-\mu}{\sigma}\right)^2\right\},\qquad(8.14)$$

and the range is the whole of $\mathbb{R}$. To show that (8.14) is a genuine pdf requires some ingenuity with double integrals and is postponed to Appendix 3.

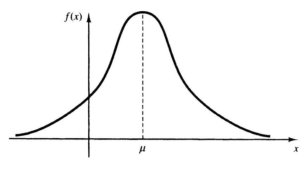

Fig. 8.8.

This pdf is widely known for its 'bell-shaped curve', as shown in Fig. 8.8. The graph indicates that the curve is symmetric about the line $y = \mu$. Note also the asymptotic behaviour of the curve in both directions along the x-axis.

In general if X has the pdf of (8.14), we say that it is a *normal (or normally distributed) random variable with parameters μ and σ^2* and we write

$$X \sim N(\mu, \sigma^2).$$

The parameters μ and σ^2 have both a probabilistic and a geometric interpretation. Probabilistically, a routine exercise in integration shows that

$$\mathbb{E}(X) = \mu \text{ and } \text{Var}(X) = \sigma^2.$$

You can establish these for yourself in Exercise 8.25 or, alternatively, wait for the next section where you'll meet a short cut. Geometrically, the story is told by Figs. 8.9 and 8.10. In Fig. 8.9 we have the pdfs for two normal random variables with the same variance but different means $\mu_1 < \mu_2$. In Fig. 8.10 the means are the same but the variances are different, with $\sigma_1 < \sigma_2$.

To calculate probabilities for normal random variables, we need the cumulative distribution function $F(x) = \int_{-\infty}^{x} f(y)\mathrm{d}y$, with f, as given by (8.14). It turns out that there is no way of expressing F in terms of elementary functions such as polynomials, exponentials or trigonometric functions (try it!). Consequently, we need to use numerical methods to calculate F. Fortunately, this has already been done for us to a considerable degree of accuracy and the results are commercially available in various published statistical tables (if you want to experiment yourself,

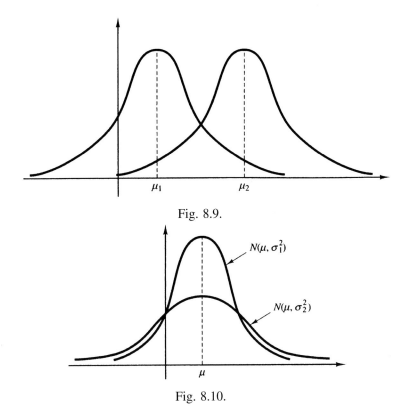

Fig. 8.9.

Fig. 8.10.

see Exercise 8.22). Of course, it would be impossible to prepare these for all of the infinite number of different normal random variables obtained for different values of μ and σ but, fortunately, we can 'standardise'. More precisely, define a new random variable Z by

$$Z = \frac{X - \mu}{\sigma}. \tag{8.15}$$

Lemma 8.5

$$Z \sim N(0, 1).$$

Proof Make the substitution $z = \frac{x-\mu}{\sigma}$ in (8.14) so that $dx = \sigma \, dz$. The cumulative distribution for Z is then given by

$$F(z) = P(Z \le z) = \frac{1}{\sqrt{(2\pi)}} \int_{-\infty}^{z} \exp\left(-\frac{1}{2}y^2\right) dy.$$

which corresponds to the pdf of (8.14) with $\mu = 0$ and $\sigma = 1$. □

The random variable Z is called the *standard normal*. All statistical tables on the market give probabilities associated with Z for values $z \ge 0$. However, some of these are in terms of the cumulative distribution $F(z) = P(Z \le z)$ (area 1 in

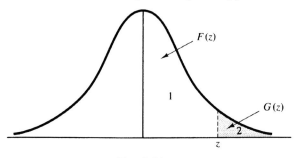

Fig. 8.11.

Fig. 8.11), while others give the so-called 'tail probability' $G(z) = 1 - F(z) = P(Z \geq z)$ (area 2 in Fig. 8.11).

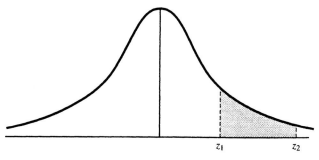

Fig. 8.12.

Note that for $z_1 < z_2$, we have (Fig. 8.12)

$$P(z_1 \leq Z \leq z_2) = F(z_2) - F(z_1)$$
$$= G(z_1) - G(z_2)$$
$$= P(Z \geq z_1) - P(Z \geq z_2). \qquad (8.16)$$

Warning: Before doing numerical exercises, be clear which type of table you have.

All the calculations in this book are carried out with the aid of tables of $G(z)$, which the reader may find in Appendix 4.

A useful property of the standard normal which gets around the lack of negative values in the tables is the following.

Lemma 8.6 *For $z \geq 0$*

$$P(Z \leq -z) = P(Z \geq z).$$

Fig. 8.13 says it all but for those who don't trust diagrams we have the following:

Proof

$$P(Z \leq -z) = \frac{1}{\sqrt{(2\pi)}} \int_\infty^{-z} \exp\left(-\frac{1}{2}y^2\right) dy.$$

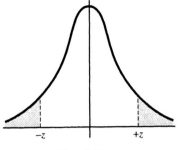

Fig. 8.13.

Now substitute $t = -y$ to find

$$P(Z \leq -z) = -\frac{1}{\sqrt{(2\pi)}} \int_{\infty}^{z} \exp\left(-\frac{1}{2}t^2\right) dt$$

$$= \frac{1}{\sqrt{(2\pi)}} \int_{z}^{\infty} \exp\left(-\frac{1}{2}t^2\right) dt = P(Z \geq z).$$

□

The trick in numerical exercises such as that given below is to manipulate probabilities using the usual laws together with properties of Z such as Lemma 8.6 above until the required probability is expressed entirely in terms of G (or F) and so can then be found by using the tables.

Example 8.6 A normally distributed random variable X has a mean of 75 and a standard deviation of 13. Find:

(a) $P(X \leq 81)$,
(b) $P(72 \leq X \leq 80)$.

Solution We will use tables which give $G(z) = P(Z \geq z)$. First note that, by Lemma 8.5, we have that $Z \sim N(0, 1)$, where

$$Z = \frac{X - 75}{13}.$$

(a) We have

$$P(X \leq 81) = P\left(Z \leq \frac{81 - 75}{13}\right) = P(Z \leq 0.462).$$

But

$$P(Z \leq 0.462) = 1 - P(Z \geq 0.462)$$

$$= 1 - G(0.462)$$

$$= 1 - 0.3228 \qquad \text{(by tables)}$$

$$= 0.68 \qquad \text{(to two decimal places)}.$$

(b)

$$P(72 \leq X \leq 80) = P\left(\frac{72-75}{13} \leq Z \leq \frac{80-75}{13}\right)$$

$$= P(-0.231 \leq Z \leq 0.385)$$

$$= P(Z \geq -0.231) - P(Z \geq 0.385) \qquad\qquad \text{by (8.15)}$$

$$= 1 - P(Z \leq -0.231) - P(Z \geq 0.385)$$

$$= 1 - P(Z \geq 0.231) - P(Z \geq 0.385) \qquad\qquad \text{by Lemma 8.6}$$

$$= 1 - G(0.231) - G(0.385)$$

$$= 0.24 \text{ to two decimal places.}$$

The importance of normal random variables derives from the fact that they describe situations where a large number of values of a random variable are clustered around the mean, with values away from the mean becoming more and more unlikely the further away we go. Random variables of this type arise in a number of important practical situations such as the following:

(a) *The theory of measurement.* Repeated experimental measurements of a given quantity appear to have the property of clustering around the 'true value' in the way described (at least approximately) by the normal curve. The variation is caused by random 'noise', which even the most careful experimental procedure is unable to eliminate. This manifestation of the normal distribution was a major impetus in the development of the central limit theorem, which is described in the next chapter.

(b) Nature provides many examples of phenomena which are modelled by normal random variables. Examples include the spread of many human attributes across the population, such as height, weight and IQ scores (whatever the merits of the latter may be as indicators of 'intelligence').

8.6 The central limit theorem

Just as the normal random variable is (probably) the most important random variable in elementary probability theory, then the central limit theorem is perhaps the most important single result. Indeed, it is a result which clarifies the role of the normal distribution and has vital implications for statistics.

Before stating and proving this theorem, we must first study the technique of moment generating functions. These were introduced for discrete random variables in Exercises 5.37–5.39. We now generalise to the continuous case.

Let X be a random variable with range (a, b) and pdf f. Its *moment generating function* M_X is defined by

$$M_X(t) = \int_a^b e^{tx} f(x) dx \tag{8.17}$$

for $t \in \mathbb{R}$.

Note: Clearly, M_X is only defined when the integral is finite. This may not always be the case and for this reason more advanced books on probability theory tend to use an object called the 'characteristic function' instead of M_X. This is defined as in (8.17) but with t replaced by it where $i = \sqrt{-1}$, and it is finite for all random variables. Knowledgeable readers will recognise (8.17) as the Laplace transform of the function f, whereas the characteristic function is the Fourier transform.

The reason why M_X is called a moment generating function is the following (see Exercise 5.37).

Lemma 8.7

$$\mathbb{E}(X^n) = \frac{d^n}{dt^n} M_X(t) \bigg|_{t=0}. \tag{8.18}$$

Proof Differentiate under the integration sign in (8.17) and use

$$\frac{d^n}{dt^n} (e^{tx}) = x^n e^{tx}.$$

Now put $t = 0$ and compare the result with (8.8). □

Another way of proving (8.18) is to expand the exponential in (8.17) as an infinite series,

$$e^{tx} = \sum_{n=0}^{\infty} \frac{t^n x^n}{n!},$$

and recognise that (8.17) then becomes

$$M_X(t) = \sum_{n=0}^{\infty} \frac{t^n}{n!} \mathbb{E}(X^n). \tag{8.19}$$

We will find (8.19) of value below.

Example 8.7 Find $M_X(t)$ for $X \sim N(\mu, \sigma^2)$. Hence, verify that the mean and variance of X are μ and σ^2 respectively.

Solution Substituting (8.14) in (8.17) we find

$$M_X(t) = \frac{1}{\sigma\sqrt{(2\pi)}} \int_{-\infty}^{\infty} e^{xt} \exp\left\{-\frac{1}{2}\left(\frac{x-\mu}{\sigma}\right)^2\right\} dx.$$

Substitute $z = \frac{x-\mu}{\sigma}$ to obtain

$$M_X(t) = \frac{1}{\sqrt{(2\pi)}} e^{\mu t} \int_{-\infty}^{\infty} e^{\sigma t z} \exp\left(-\frac{1}{2}z^2\right) dz$$

$$= \frac{1}{\sqrt{(2\pi)}} \exp\left(\mu t + \frac{1}{2}\sigma^2 t\right) \int_{\infty}^{\infty} \exp\left(-\frac{1}{2}(z - \sigma t)^2\right) dz.$$

Now substitute $y = z - \sigma t$ and use the result of Appendix 3 to find

$$M_X(t) = \exp\left(\mu t + \frac{1}{2}\sigma^2 t^2\right). \tag{8.20}$$

Notice, in particular, that if $Z \sim N(0, 1)$, then

$$M_Z(t) = \exp\left(\frac{1}{2}t^2\right).$$

Differentiate (8.20) twice and use (8.18) to find $\mathbb{E}(X) = \mu$ and $\mathbb{E}(X^2) = \sigma^2 + \mu^2$. Hence, $\text{Var}(X) = \sigma^2$ by Theorem 5.6(a) (extended via Exercise 8.14).

Now let $X_1, X_2, \ldots$ be a sequence of i.i.d. random variables, each with mean μ and variance σ^2, and consider their sum $S(n)$, which, we recall, has mean $n\mu$ and variance $n\sigma^2$. We note that the results of Exercise 5.39(a) and (b) extend to our more general situation and we will feel free to use these below (full proofs will be given in Section 9.5). We standardise the sum to form the random variables

$$Y(n) = \frac{S(n) - n\mu}{\sigma\sqrt{n}} \tag{8.21}$$

and note that $\mathbb{E}(Y(n)) = 0$ and $\text{Var}(Y(n)) = 1$.

We are now ready to state and prove the central limit theorem. Before we do this let's take a quick peek forward and find out what it's going to tell us. The startling news is that whatever the nature of the random variables X_j – whether they are discrete or continuous, Bernoulli or exponential – as n gets larger and larger the distribution of $Y(n)$ always gets closer and closer to that of the standard normal Z! We will explore some of the consequences of this astonishing result after the proof.

Theorem 8.8 *(the central limit theorem) Let $X_1, X_2, \ldots$ be i.i.d. random variables with common mean μ and variance σ^2. Write $S(n) = X_1 + X_2 + \cdots + X_n$; then for any $-\infty \le a < b \le \infty$, we have*

$$\lim_{n \to \infty} P\left(a \le \frac{S(n) - n\mu}{\sigma\sqrt{n}} \le b\right) = \frac{1}{\sqrt{(2\pi)}} \int_a^b \exp\left(-\frac{1}{2}z^2\right) dz. \tag{8.22}$$

Proof We will only give a brief outline of the proof which makes it appear to be a lot simpler than it really is! Our strategy will be to use moment generating functions and we will, at the end, use the following fact which is too difficult to prove herein.

If $W_1, W_2, \ldots$ is a family of random variables such that for all $t \in \mathbb{R}$, $\lim_{n \to \infty} M_{W_n}(t) = M_W(t)$, where M_W is itself the moment generating function of a random variable W, then

$$\lim_{n \to \infty} P(a \le W_n \le b) = P(a \le W \le b) \tag{#}$$

for all $-\infty \le a < b \le \infty$.

Now consider i.i.d. random variables defined by

$$T_j = X_j - \mu \quad \text{for } j = 1, 2, \ldots.$$

Clearly, we have $\mathbb{E}(T_j) = 0$ and $\mathbb{E}(T_j^2) = \text{Var}(T_j) = \sigma^2$ and

$$Y(n) = \frac{1}{\sigma \sqrt{n}} (T_1 + T_2 + \cdots + T_n)$$

where $Y(n)$ is given by (8.21).

Now by Exercise 8.32 with $a = \sigma \sqrt{n}$ and $b = 0$ and Exercise 5.39(b) we have

$$M_{Y(n)}(t) = \left(M \left(\frac{t}{\sigma \sqrt{n}} \right) \right)^n \quad \text{for all } t \in R$$

where M on the right-hand side denotes the common moment generating function of the T_is.

Now apply (8.19) to M to find

$$M \left(\frac{t}{\sigma \sqrt{n}} \right) = 1 + \left(\frac{t}{\sigma \sqrt{n}} \times 0 \right) + \left(\frac{1}{2} \frac{t^2}{\sigma^2 n} \sigma^2 \right) + \sum_{m=3}^{\infty} n^{-m/2} \frac{t^m}{m!} \frac{\mathbb{E}(T_j^m)}{\sigma^m}$$

$$= 1 + \frac{1}{2} \frac{t^2}{n} + \left(\frac{1}{n} \times \alpha(n) \right)$$

where $\alpha(n) = \sum_{m=3}^{\infty} n^{-(m-2)/2} \frac{t^m}{m!} \frac{\mathbb{E}(T_j^m)}{\sigma^m}$ and we note that $\mathbb{E}(T_j^m)$ has the same value for any $j = 1, 2, \ldots$. Thus

$$M_{Y(n)}(t) = \left(1 + \frac{1}{2} \frac{t^2}{n} + \frac{1}{n} \alpha(n) \right)^n.$$

Now it is well known that $e^y = \lim_{n \to \infty} (1 + \frac{y}{n})^n$ and, furthermore, if $\lim_{n \to \infty} \alpha(n) = 0$, then it can be shown that

$$e^y = \lim_{n \to \infty} \left(1 + \frac{y}{n} + \frac{1}{n} \alpha(n) \right)^n$$

(See Exercise 8.33 if you want to prove this for yourself.)

Hence we find that

$$\lim_{n\to\infty} M_{Y(n)}(t) = \exp\left(\frac{1}{2}t^2\right) = M_Z(t)$$

and the final result then follows from (#) above. □

As a first application of this result we can begin to understand the so-called 'binomial approximation to the normal distribution', which we hinted at in Section 8.4. Indeed, if we take $X_1, X_2, \ldots$ to be Bernoulli random variables with common parameter p, then $S(n)$ is binomial with parameters n and p, and (8.22), together with (5.11), yields the following corollary.

Corollary 8.9 *(de Moivre–Laplace central limit theorem) If $X_1, X_2, \ldots$ are Bernoulli with common parameter p, then*

$$\lim_{n\to\infty} P\left(a \le \frac{S(n) - np}{\sqrt{[np(1-p)]}} \le b\right) = \frac{1}{\sqrt{2\pi}} \int_a^b \exp\left(-\frac{1}{2}z^2\right) dz \qquad (8.23)$$

Equation (8.23) was historically the first central limit theorem to be established. It is named after de Moivre who first used integrals of $\exp(-\frac{1}{2}z^2)$ to approximate probabilities, and Laplace who obtained a formula quite close in spirit to (8.23). It is fascinating that Laplace was motivated in these calculations by the need to estimate the probability that the inclination of cometary orbits to a given plane was between certain limits. The modern central limit theorem (8.22) began to emerge from the work of the Russian school of probabilists – notably Chebyshev and his students Markov and Lyapunov towards the close of the nineteenth century.

Example 8.8 A large file containing 1000 symbols is to be sent over a modem between two PCs. Owing to a fault on the line, a large number of errors is expected; however, previous experience leads us to believe that errors are probabilistically independent and occur with probability 0.05. Estimate the probability that there are more than 60 errors.

Solution We number the symbols in order of reception from 1 to 1000 and introduce the Bernoulli random variables

$$X_n = 1 \text{ if the } n\text{th symbol is an error,}$$
$$X_n = 0 \text{ if the } n\text{th symbol is not an error,}$$

for $1 \le n \le 1000$, and note that

$$P(X_n = 1) = 0.05.$$

We are interested in the distribution of $S(1000) = X_1 + X_2 + \cdots + X_{1000}$ and note that this has a binomial distribution with parameters $n = 1000$ and $p = 0.05$. It follows that it has mean

$$\mu = 0.05 \times 1000 = 50$$

and standard deviation

$$\sigma = \sqrt{(1000 \times 0.05 \times 0.95)} = 6.89.$$

We require

$$P(S(1000) \geq 60).$$

To calculate this directly using the binomial distribution would be very long and tedious (unless you have access to a program which can do it for you). We use the de Moivre–Laplace central limit theorem to obtain a good approximation, that is we approximate $\frac{S(1000)-50}{6.89}$ by the standard normal Z to find

$$P(S(1000) \geq 60) = P\left(Z \geq \frac{60-50}{6.89}\right) = G(1.45)$$

$$= 0.074 \text{ by tables.}$$

The statistical applications of the central limit theorem follow from dividing the top and bottom of (8.21) by n and recalling that the values of the random variable $\overline{X}(n)$ are all possible sample means calculated from samples of size n. Equation (8.22) then takes the form

$$\lim_{n \to \infty} P\left(a \leq \frac{\overline{X}(n) - \mu}{\frac{\sigma}{\sqrt{n}}} \leq b\right) = \frac{1}{\sqrt{(2\pi)}} \int_a^b \exp\left(-\frac{1}{2}z^2\right) dz. \qquad (8.24)$$

In practical statistical calculations it is often a reasonable approximation to treat $\frac{\overline{X}(n)-\mu}{\sigma/\sqrt{n}}$ as a standard normal when $n \geq 20$. This is the basis of much elementary work on confidence intervals and hypothesis tests. Equation (8.23) in particular forms the basis for statistical tests on proportions.

We invite the reader to compare what the law of large numbers and the central limit theorem are saying about the asymptotic behaviour of $\overline{X}(n)$. In fact, the central limit theorem is giving far more information about the distribution of $\overline{X}(n)$, but this is consistent with the weak (and strong) laws in that as n gets larger and larger the standard deviation $\frac{\sigma}{\sqrt{n}}$ is getting smaller and smaller so that the pdf for $\overline{X}(n)$ is becoming more and more highly concentrated around the mean (Fig. 8.14).

Example 8.9 [the Brownian motion process] As a final application of the central limit theorem we consider the problem of how to model Brownian motion. This is the phenomenon, first discovered by the botanist Robert Brown in the nineteenth century, whereby a grain of pollen in a solution of water appears to dance and jiggle around in a random manner. The problem of trying to explain this from a physical point of view attracted the attention of a number of thinkers (including Albert Einstein). The modern explanation for Brownian motion is that the motion is caused by the bombardment of the pollen grain by the molecules of the water. We are interested in trying to model the displacement of the pollen grain at time t.

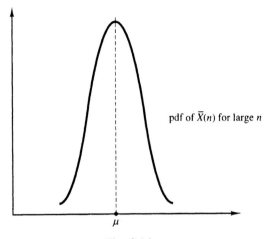

μ

Fig. 8.14.

As its motion is random, we denote this displacement by a random variable $X(t)$. So the path followed by the pollen grain is described by the family of random variables $(X(t), t \geq 0)$, all defined on the same probability space (such a family of random variables is usually called a 'stochastic process' [*]). To simplify matters we will assume that the displacement is taking place in one spatial dimension only.

Observation tells us that 'on average' the pollen grains don't seem to get anywhere so we take $\mathbb{E}(X(t)) = 0$. Furthermore, if $0 \leq t_1 \leq t_2$, then the motion between t_1 and t_2 doesn't look any different than that between 0 and t_1. This suggests in particular that the random variables $X(t_1)$ and $X(t_2) - X(t_1)$ should be independent and that the distribution of $X(t_2) - X(t_1)$ should depend only upon the time interval $t_2 - t_1$.

If we write $V(t) = \text{Var}(X(t))$, we then find by a suitable extension of Theorem 5.11(c) (see Exercise 9.16) that

$$V(t_2) = V(t_1) + V(t_2 - t_1)$$

which is a linear equation and has the solution $V(t) = At$, where A is a constant.

We would like to know the distribution of each $X(t)$. We argue as follows: each $X(t)$ arises as the result of bombardment by a large number of identical water molecules (experiment indicates that there are around 10^{21} of these per second) (Fig. 8.15). It seems reasonable to model the molecular impacts as i.i.d. random variables.

The central limit theorem then tells us that (irrespective of the precise nature of the molecular impacts), as $X(t)$ is the sum of a large number of i.i.d. random

---

[*] You can learn more about this important concept in Chapter 10.

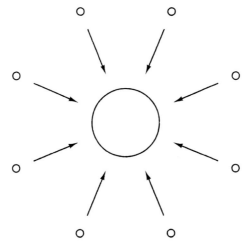

Fig. 8.15.   Bombardment of pollen grain by water molecules.

variables, then $\frac{X(t)}{\sqrt{(At)}}$ should be well-approximated by a standard normal. It then follows by Exercise 8.28 that

$$X(t) \sim N(0, At).$$

Notes

(i) The constant A is related to Avogadro's constant. Perrin won the Nobel Prize in 1926 for an experimental determination of Avogadro's constant based on Brownian motion.

(ii) The stochastic process called Brownian motion described above is now known not to be the best mathematical model available to describe the physical phenomenon of Brownian motion. Nonetheless, its properties have been, and are, continuing to be relentlessly explored in the twentieth and twenty-first centuries, and it has become an essential ingredient of mathematical models of a large number of situations where random noise is involved, from systems of diffusing particles to the mysterious workings of the Stock Exchange.

8.7 Entropy in the continuous case

We want to generalise some of the ideas of Chapter 6 to the case where X is a random variable taking values in some interval $[a, b]$. The motivation for this is the need to develop a theory of information which is appropriate for the modelling of the transmission of continuous signals. In this section we will consider how the concept of entropy carries over to the more general case. We begin with a definition.

Let X be a random variable with range (a, b) and pdf f. We define its *entropy* $H(X)$ to be

$$H(X) = -\int_a^b \log(f(x))f(x)dx$$

$$= \mathbb{E}\left(\log\left(\frac{1}{f(X)}\right)\right), \tag{8.25}$$

where we recall the convention from Chapter 6 that log is the logarithm to base 2.

Example 8.10 Calculate the entropy for $X \sim U(a, b)$.

Solution We recall that $f(x) = \frac{1}{b-a}$ for $x \in (a, b)$; hence, by (8.25), we have

$$H(X) = -\int_a^b \log\left(\frac{1}{b-a}\right)\frac{1}{b-a}dx$$

$$= \log(b - a).$$

When we discussed entropy for discrete random variables in Chapter 6, we found that it had a number of properties that made it a natural measure of uncertainty; in particular, we always had $H(X) \geq 0$. This latter condition is violated in the general case as we see by taking $X \sim U(a, b)$ with a and b such that $0 < b - a < 1$ (e.g. $a = 0.1$ and $b = 0.5$).

The above suggests that entropy in the continuous case is not such a natural concept as in the discrete case. More evidence for this proposition is obtained by going through the procedure of discretisation and integration described in Section 8.3. We define a partition $\mathcal{P}$ such that $a = x_0 < x_1 < \cdots < x_{n-1} < x_n = b$ and choose $\xi_j \in (x_{j-1}, x_j)$ for $1 \leq j \leq n$. The discretisation $\hat{X}$ of X then has the law

$$P(\hat{X} = \xi_j) = p_j, \text{ where } p_j = \int_{x_{j-1}}^{x_j} f(x)dx$$

for $1 \leq j \leq n$.

By (6.3), we have

$$H(\hat{X}) = -\sum_{j=1}^n p_j \log(p_j),$$

and we want to see what happens to this in the limit. Now

$$H(\hat{X}) = -\sum_{j=1}^n p_j \log(f(\xi_j)) + \sum_{j=1}^n p_j \log\left(\frac{f(\xi_j)}{p_j}\right).$$

Now before we take the limit we use the same trick as in Section 8.3 and replace each p_j by $\frac{p_j}{\Delta x_j} \Delta x_j$. When we do this we find that

$$\lim \left(-\sum_{j=1}^{n} p_j \log(f(\xi_j)) \right) = H(X),$$

as given by (8.25) above, but there is a problem in the second term in $H(\hat{X})$ which contains within it a sum of 'nicely behaved terms' multiplied by $-\log(\Delta x_j)$, which tends to $+\infty$ in the limit. Thus we find that

$$\lim H(\hat{X}) = H(X) + \infty = \infty.$$

This suggests that there is no adequate version of entropy in the continuous case. In extracting the term $H(X)$ from the infinite limit, we are in a sense carrying out a procedure which physicists call 'renormalisation' (to hide the fact that they are on dodgy ground).

We should not expect entropy to play as big a role in the continuous case as it does in the discrete case. Fortunately, the concept of mutual information passes over nicely through the process of discretisation and integration, as we will see in the next chapter. We will also find that entropy does have its uses in helping us to calculate channel capacities, and the following results will be of use there.

Example 8.11 Find the entropy of $X \sim N(\mu, \sigma^2)$.

$$H(X) = -\frac{1}{\sigma(2\pi)^{1/2}} \int_{-\infty}^{\infty} \exp\left(-\frac{1}{2}\left[\frac{x-\mu}{\sigma} \right]^2 \right)$$

$$\times \log\left\{ \frac{1}{\sigma(2\pi)^{1/2}} \exp\left(-\frac{1}{2}\left[\frac{x-\mu}{\sigma} \right]^2 \right) \right\} dx$$

$$= \log(\sigma(2\pi)^{1/2}) + \frac{\log(e)}{\pi^{1/2}} \int_{-\infty}^{\infty} e^{-y^2} y^2 dy,$$

where we have used the substitution $y = \frac{x-\mu}{2^{1/2}\sigma}$.

Now use the fact that

$$\int_{-\infty}^{\infty} e^{-y^2} y^2 dy = \frac{1}{2}\pi^{1/2} \text{(as Var}(Z) = 1 \text{ for } Z \sim N(0, 1))$$

to find

$$H(X) = \log(\sigma(2\pi)^{1/2}) + \frac{\log(e)}{\pi^{1/2}} \frac{1}{2}\pi^{1/2},$$

that is,

$$H(X) = \log(\sigma(2\pi e)^{1/2}). \tag{8.26}$$

We will in the sequel write $H_N(\sigma)$ for the quantity given by (8.26). It is interesting to observe that H_N is a function of σ but not of μ – this is an indication of the strong relationship between variance and uncertainty.

Now let S denote the set of all random variables with range $\mathbb{R}$ which possess a pdf and have mean μ and variance σ^2.

Theorem 8.10 *If $X \in S$, then*

$$H(X) \leq H_N$$

with equality if and only if $X \sim N(\mu, \sigma^2)$.

Proof We adopt the elegant proof given by Goldie and Pinch which uses the continuous version of the Gibbs inequality – you can prove this for yourself in Exercise 8.38. This states that, with $H(X)$ given as above by (8.25), we have for any pdf g

$$H(X) \leq -\int_a^b \log(g(x)) f(x) dx$$

with equality if and only if $f = g$.

Now take $X \in S$ with pdf f and take g to be the normal distribution (8.14); then we find that

$$-\int_{-\infty}^{\infty} \log(g(x)) f(x) dx = \log(\sigma (2\pi)^{1/2}) + \frac{\log(e)}{2\sigma^2} \int_{-\infty}^{\infty} (y - \mu)^2 f(y) dy$$

$$= \frac{1}{2} \log(2\pi\sigma^2) + \frac{\log(e)}{2\sigma^2} \operatorname{Var}(Y).$$

But $\operatorname{Var}(Y) = \sigma^2$, hence by the Gibbs inequality we have

$$H(X) \leq \frac{1}{2} \log(2\pi e \sigma^2)$$

as is required. The equality part is immediate. □

Exercises

8.1. For a random variable X which is distributed continuously on (a, b) show that for $[c, d] \subset (a, b)$ we have

$$p_X([c, d]) = p_X([c, d)) = p_X((c, d]) = p_X((c, d)).$$

8.2. A random variable X has range (1,8) and law given by

$$p_X(\{1\}) = \frac{1}{2}, \quad p_X(\{2\}) = \frac{1}{4} \quad \text{and} \quad p_X((c, d)) = \frac{1}{120} \int_c^d x dx$$

whenever $(c, d) \subseteq (2, 8)$. Confirm that p_X really is a probability measure. (Note that X is neither discrete nor continuously distributed.)

8.3. Establish the following continuous version of lemmatta 5.2–5.3 for X a random variable with range (a, b) and $a \le c < d \le b$:

   (i) F is an increasing function of x, that is, $F(c) \le F(d)$.

   (ii) $F(a) = 0$, $F(b) = 1$.

   (iii) $P(X \ge c) = 1 - F(c)$.

8.4. A random variable X has range $(0,2)$ and cumulative distribution

$$F(x) = \frac{1}{20}(x + 4)^2 - \frac{4}{5}.$$

   (a) Find the pdf for X.

   (b) Calculate $P(0 \le X \le 1)$.

8.5. Find the constant $c \in \mathbb{R}$ such that the following functions are pdfs on the given interval:

   (i) $\frac{3}{56}(x^2 + 2x + c)$ on $(1, 3)$,

   (ii) $c \cos(\pi x)$ on $(0, \frac{1}{2})$,

   (iii) $\frac{1}{x-4}$ on $(5, c)$.

8.6. A random variable X has pdf

$$f(x) = \frac{3}{88}x(x + 1) \text{ on } (0, 4).$$

   (i) Confirm that f really is a pdf.

   (ii) Compute the cumulative frequency distribution of X.

   (iii) Calculate $P(2 \le X \le 3)$.

8.7. Let $X \sim U(a, b)$ with pdf g and $Y \sim U(0, 1)$ with pdf f. Show that

$$g(x) = \frac{1}{b - a} f\left(\frac{x - a}{b - a}\right) \quad \text{for } x \in (a, b).$$

8.8. (i) Find the mean and variance of $X \sim U(a, b)$.

   (ii) Show that

$$\mathbb{E}(X^n) = \frac{1}{n + 1}(b^n + ab^{n-1} + a^2 b^{n-2} + \cdots + a^{n-1}b + a^n).$$

8.9. Find the variance of $X \sim \mathcal{E}(\lambda)$.

8.10. Find the constant c such that

$$f(x) = \frac{c}{1 + x^2}$$

is a pdf on $\mathbb{R}$. The resulting random variable is said to be Cauchy. Show that its mean is not finite.

8.11. A gamma random variable X with parameters $\alpha > 0$ and $\beta > 0$ has range $[0, \infty)$ and pdf

$$f(x) = \frac{x^{\alpha-1} \exp(-x/\beta)}{\Gamma(\alpha)\beta^\alpha}$$

where Γ is the gamma function described in Section 2.5:

(i) Confirm that f is a pdf.
(ii) Check that when $\alpha = 1$, $X \sim \mathcal{E}(1/\beta)$.
(iii) Deduce that

$$\mathbb{E}(X^n) = \beta^n \frac{\Gamma(n + \alpha)}{\Gamma(\alpha)}.$$

(iv) Use the result of (iii) to find $\mathbb{E}(X)$ and $\mathrm{Var}(X)$.

(*Note*: Gamma random variables give a more sophisticated approach to modelling the lifetimes of some components than the exponential random variables. A special case of some interest is the χ^2 (chi-squared) random variable with n degrees of freedom whose pdf is obtained by taking $\alpha = n/2$ and $\beta = 2$. This latter random variable is used for statistical tests concerning the variance of populations.)

8.12. Another class of random variables which is useful in modelling lifetimes is the Weibull class. This has range $[0, \infty)$, parameters $\gamma > 0$ and $\theta > 0$ and pdf

$$f(x) = \frac{\gamma}{\theta} x^{\gamma-1} \exp\left(-\frac{x^\gamma}{\theta}\right).$$

(i) Confirm that this yields a pdf.
(ii) Find the value of γ for which $X \sim \mathcal{E}(1/\theta)$.
(iii) Find $\mathbb{E}(X)$ and deduce the form of $\mathbb{E}(X^n)$.

8.13. If X is a random variable with range (a, b) and Y is a random variable with range (c, d), write down the ranges of:

(a) $X + Y$,
(b) αX,
(c) XY.

8.14. Extend Theorems 5.5 and 5.6 to the continuous case.
8.15. In the proof of Chebyshev's inequality, show that $A = (-\infty, \mu - c] \cup [\mu + c, \infty)$. What is $\overline{A}$?
8.16. Prove Chebyshev's inequality for discrete random variables.
8.17. Why can't the strong law of large numbers be used to 'define' probability as a limit of relative frequencies?
8.18. Show that if X and Y are identically distributed, then:

(i) X is continuously distributed if and only if Y is,
(ii) X has pdf f if and only if Y has.

8.19. A random variable X has mean 2 and variance 1.6. Estimate the probability that X is greater than 5.
8.20. Prove Markov's inequality, that is if X is a random variable with range $[0, \infty)$, then

$$P(X > \lambda) \le \frac{\mathbb{E}(X)}{\lambda}.$$

[*Hint*: Proceed as in the proof of Chebyshev's inequality but use $\mathbb{E}(X)$ instead of $\mathrm{Var}(X)$.]

8.21. If $X \sim N(0, 1)$, use tables to find:

    (a) $P(Z > 1.34)$,
    (b) $P(Z < 1.34)$,
    (c) $P(Z < -1.34)$,
    (d) $P(Z > -1.34)$,
    (e) $P(0.68 < Z < 2.17)$,
    (f) $P(-1.76 < Z < 0.24)$,
    (g) $P(Z = 0.43)$.

8.22. Use Simpson's rule with eight strips (or an alternative method of numerical integration if you prefer) to estimate $P(0 < Z < 1)$ and compare your result with that given in the tables.

8.23. Without using tables, show that

$$P(Z \le 0) = P(Z \ge 0) = \frac{1}{2}.$$

If $X \sim N(\mu, \sigma^2)$, what are $P(X \ge \mu)$ and $P(X \le \mu)$?

8.24. If $Z \sim N(0, 1)$ and $a > 0$, show that

$$P(Z > -a) = 1 - P(Z > a)$$

and hence deduce that

$$P(-a < Z < a) = 1 - 2P(Z > a).$$

8.25. If $X \sim N(\mu, \sigma^2)$, show that

$$\mathbb{E}(X) = \mu \text{ and } \mathrm{Var}(X) = \sigma^2.$$

[*Hint*: First show that $E(Z) = 0$ and $\mathrm{Var}(Z) = 1$.]

8.26. A continuous signal is transmitted over a binary symmetric channel. The number of errors per second is found to be approximately normal with mean 2 and variance 0.75. Find the probability of (a) less than one error, (b) between one and three errors, (c) more than four errors, per second.

8.27. A random variable is normally distributed and has variance 1.6. It is known that $P(X \ge 1.8) = 0.1$. Find the mean of X.

8.28. If $X \sim N(\mu, \sigma^2)$, show that $Y = cX + d$ is also normally distributed (where c and $d \in \mathbb{R}$). Find the mean and variance of Y.

8.29.* Prove the tail inequality for $Z \sim N(0, 1)$, that is that

$$G(y) \le \exp\left(-\frac{1}{2}y^2\right) \quad \text{for } y \ge 0.$$

[*Hint*: Use the fact that $e^{y(z-y)} \ge 1$ whenever $z \ge y$.]

8.30. Find the moment generating function for:

    (i) $X \sim U(a, b)$,
    (ii) $X \sim \mathcal{E}(\lambda)$.

Use the result of (ii) to find the nth moment for $X \sim \mathcal{E}(\lambda)$.

8.31. Find the moment generating function for the gamma random variable of Exercise 8.11.

8.32. Show that if M_X is the moment generating function for a random variable X with range $\mathbb{R}$, then $aX + b$ has moment generating function given by

$$M_{aX+b}(t) = e^{bt} M_X(at).$$

8.33.* Convince yourself that

$$\lim_{n \to \infty} \left(1 + \frac{y + \alpha(n)}{n} \right)^n = e^y \quad \text{if} \quad \lim_{n \to \infty} \alpha(n) = 0.$$

8.34. A coin is biased so that the probability of heads is 0.45. Estimate the probability that, after 25 throws, the coin has come up heads between 45% and 55% of the time.

8.35. Estimate the number of times that a fair coin should be tossed to ensure that the probability that between 48% and 52% of the results are heads is 0.99.

8.36. Write down the pdf as a function $p(x, t)$ of position and time for the random variable $X(t)$ describing the displacement after t seconds of a particle undergoing Brownian motion. Show that p satisfies the diffusion equation

$$\frac{\partial p}{\partial t} = \frac{A}{2} \frac{\partial^2 p}{\partial x^2}.$$

8.37. Find the entropy of $X \sim \mathcal{E}(\lambda)$.

8.38. Imitate the procedure of Exercise 6.9 to prove the Gibbs inequality in the continuous case, that is that

$$-\int_a^b f(x) \log(f(x)) \mathrm{d}x \leq -\int_a^b f(x) \log(g(x)) \mathrm{d}x$$

with equality if and only if $f = g$.

8.39. Use the result of Exercise 8.38 to show that if X is a random variable with pdf f, then:

(a) if X has finite range (a, b), then X has maximum entropy when $X \sim U(a, b)$,

(b) if X has range $[0, \infty)$ and expectation $\mathbb{E}(X) = \frac{1}{\lambda}$, where $\lambda > 0$, then X has maximum entropy when $X \sim \mathcal{E}(\lambda)$.

8.40. A random variable X with range $[0, \infty)$ is said to be memoryless if

$$P_{(X>s)}(X > t) = P(X > t - s) \quad \text{for all } 0 \leq s < t.$$

(a) Show that if X is memoryless, then

$$P(X > t + h) = P(X > t)P(X > h) \quad \text{for all } h, t > 0.$$

(b) Show that if $X \sim \mathcal{E}(\lambda)$ for some $\lambda > 0$, then X is memoryless (in fact all memoryless random variables are of this type).

Further reading

All the books listed at the end of Chapter 5 are relevant for the material on continuous probability. Readers should be aware that both the law of large numbers and the central limit theorem can be generalised considerably. A beautiful book, written by one of the leading figures of twentieth-century probability, which extends a number of ideas from this chapter is *Introduction to Probability Theory* by K. Itô (Cambridge University Press, 1978). In particular, this contains a proof of the convergence result (Lévy convergence theorem) denoted (#) at the beginning of our 'proof' of the central limit theorem. A fascinating account of the history of the central limit theorem can be found in *The Life and Times of the Central Limit Theorem* by W. J. Adams (Kaedmon Publishing Company, 1974).

Many of the books described at the end of Chapter 7 discuss continuous entropy. I found Reza, and Goldie and Pinch particularly useful. *Information Theory for Continuous Systems* by S. Ihara (World Scientific, 1993) is a fascinating book devoted entirely to continuous information.

9

Random vectors

9.1 Cartesian products

In this chapter, we aim to study probabilities associated with events occurring in two-, or higher, dimensional spaces. Before we begin this investigation, we need to extend some of the ideas of Chapter 3 to this context.

Let A and B be sets. The *Cartesian product* of A and B is defined to be the set $A \times B$ of all ordered pairs

$$A \times B = \{(a, b),\ a \in A,\ b \in B\}.$$

The Cartesian product is named after the great seventeenth-century mathematician and philosopher René Descartes (1596–1650) (of 'I think therefore I am' fame), who effectively invented co-ordinate geometry. Sometimes we will just refer to it as a product. An example should clarify our ideas.

Example 9.1 Suppose that a fair coin is tossed, following which a fair die is rolled. What is the sample space of the combined experience?

Solution We have $A = \{H, T\}$ and $B = \{1, 2, 3, 4, 5, 6\}$. The sample space of the combined experience is then

$$A \times B = \{(H, 1),\ (H, 2),\ (H, 3),\ (H, 4),\ (H, 5),\ (H, 6),$$
$$(T, 1),\ (T, 2),\ (T, 3),\ (T, 4),\ (T, 5),\ (T, 6)\}.$$

Example 9.1 should convince the reader that, in general

$$A \times B \neq B \times A.$$

One case in which equality clearly holds is when we take $A = B$. In this case we adopt the notation

$$A^2 = A \times A.$$

An important example which will concern us much in this chapter is obtained by taking $A = \mathbb{R}$ and considering $\mathbb{R}^2 = \mathbb{R} \times \mathbb{R}$. This is nothing but the infinite plane as drawn in Fig. 9.1, and each member $(a, b) \in \mathbb{R}^2$ represents a point in the plane with x-coordinate a and y-coordinate b.

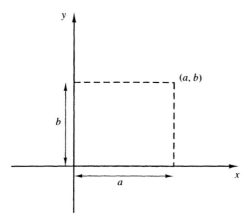

Fig. 9.1.

In general, subsets of $A \times B$ are said to be *products* if they can be written in the form

$$A_1 \times B_1 = \{(a, b) \in A \times B; a \in A_1 \subseteq A, \ b \in B_1 \subseteq B\}.$$

For example, in Example 9.1 the subset consisting of the elements $\{(H, 1), (H, 2), (H, 3)\}$ is clearly the product $\{H\} \times \{1, 2, 3\}$ and represents the event that a head is thrown followed by a 1, 2 or 3 on the die. It is not the case that every subset of a Cartesian product is a product; for example, in Example 9.1 $\{(H, 1), (T, 2)\}$ is clearly not a product. However, as we will see later, many useful subsets can be built from products.

Now we consider $\mathbb{R}^2$. One important class of examples of products consists of the *open rectangles*, which are defined to be the sets $(a, b) \times (c, d)$, where (a, b) and (c, d) are open intervals in $\mathbb{R}$ (Fig. 9.2).

Closed rectangles are defined similarly. You should convince yourself that the union of two open rectangles cannot, in general, be written as a product. Another example of a set that is not a product is the disc of radius $a > 0$, $D_a = \{(x, y) \in \mathbb{R}^2; \ x^2 + y^2 \le a^2\}$; this is shaded in Fig. 9.3. Note that the boundary of the disc is the circle of radius a defined as

$$S_a^1 = \{(x, y) \in \mathbb{R}^2; \ x^2 + y^2 = a^2\}.$$

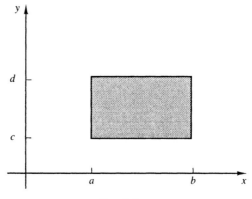

Fig. 9.2.

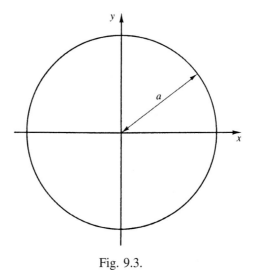

Fig. 9.3.

The above ideas generalise to define the Cartesian product of n sets $A_1, A_2, \ldots, A_n$. This is defined to be

$$A_1 \times A_2 \times \cdots \times A_n = \{(a_1, a_2, \ldots, a_n); \; a_j \in A_j, \; 1 \leq j \leq n\}.$$

A more concise notation is to denote this as $X_{j=1}^n A_j$. In particular, we write $X_{j=1}^n A = A^n$. When $n = 3$ and $A = \mathbb{R}$ we obtain the familiar three-dimensional space in which our world appears to present itself. Another relevant example is obtained by considering an experiment which is carried out n times in succession under identical circumstances. If the sample space for the first trial is S, then that for the n successive trials is S^n. A typical member of S^n represents a sequence of possible results, so the n-tuple $(x_1, x_2, \ldots, x_n)$, where each $x_j \in S$, represents the situation where the first trial gave a result of x_1, the second trial gave a result of x_2, etc.

9.2 Boolean algebras and measures on products

Let A and B be sets as above and form their product $A \times B$. We define the *product Boolean algebra* $\mathcal{B}(A) \times \mathcal{B}(B)$ to be the smallest Boolean algebra which contains all product sets $A_1 \times B_1$, where $A_1 \subseteq A$ and $B_1 \subseteq B$. As an example, recall that $\mathcal{P}(A)$, the power set of A comprising all its subsets, is a Boolean algebra. In Exercise 9.4, you can convince yourself that if A and B are finite sets, then

$$\mathcal{P}(A \times B) = \mathcal{P}(A) \times \mathcal{P}(B).$$

In the case of $\mathbb{R}$, we have already seen that $\mathcal{P}(\mathbb{R})$ is too big for many of our needs and that a more convenient Boolean algebra is that given by the broken lines $\mathcal{I}(\mathbb{R})$. For problems involving probabilities in two dimensions, we define the Boolean algebra $\mathcal{I}(\mathbb{R}^2)$ to be $\mathcal{I}(\mathbb{R}) \times \mathcal{I}(\mathbb{R})$. So $\mathcal{I}(\mathbb{R}^2)$ is the smallest Boolean algebra containing all products of broken lines. Don't worry too much about trying to visualise the most general element of $\mathcal{I}(\mathbb{R}^2)$ – it isn't necessary. However, you should convince yourself that $\mathcal{I}(\mathbb{R}^2)$ contains all the open rectangles.

Sometimes we may want to restrict ourselves to a particular region, say a rectangle $(a, b) \times (c, d)$, in which all the events of interest are taking place. In this case the relevant Boolean algebra will be $\mathcal{I}(a, b) \times \mathcal{I}(c, d)$. However, in two dimensions there are more interesting possibilities where the product Boolean algebra is not appropriate; for example, we may have a problem which exhibits circular symmetry. As an example suppose that all events take place in the disc D_a. In this case the relevant Boolean algebra may be that generated by all the subdiscs D_b where $0 < b \le a$. The following example will clarify this point.

Example 9.2 A dart is thrown randomly at a circular dartboard of radius 20 cm. Assuming that it lands somewhere on the board, what is the probability that it is within a radius of 8 cm and 12 cm from the centre (Fig. 9.4)?

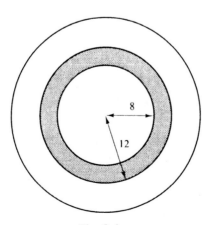

Fig. 9.4.

Solution We want $P(D_{12} - D_8)$. Noting that $D_{12} - D_8$ is in our Boolean algebra, an (obvious?) extension of the principle of symmetry leads us to

$$P(D_{12} - D_8) = \frac{\text{area of } (D_{12} - D_8)}{\text{area of } D_{20}}$$

$$= \frac{\pi (12)^2 - \pi (8)^2}{\pi (20)^2} = 0.2.$$

Although we should bear in mind the lesson of Example 9.2, there are many situations where product Boolean algebras are the natural homes for our events and we should educate ourselves about the kinds of measure that we can have for them.

If A and B are finite sets and $C \in \mathcal{B}(A) \times \mathcal{B}(B)$, then the counting measure is of course given by

$$\#C = \text{number of members of } C.$$

In particular, if $C = A_1 \times B_1$ is a product, you can easily establish (Exercise 9.2) that

$$\#(A_1 \times B_1) = (\#A_1)(\#B_1).$$

Now let A and B be arbitrary sets and let m and n be measures on the Boolean algebras $\mathcal{B}(A)$ and $\mathcal{B}(B)$ respectively. A measure μ on $B(A) \times B(B)$ is said to be the *product* of the measures m and n if, for all $A_1 \subseteq A$ and $B_1 \subseteq B$, we have

$$\mu(A_1 \times B_1) = m(A_1)n(B_1). \tag{9.1}$$

We then write $\mu = m \times n$.

Clearly, the counting measure on $A \times B$ as above is the product of the respective counting measures on A and B.

From now on we will concentrate on measures of subsets of $\mathbb{R}^2$. In particular, let m and n each be measures on $\mathcal{I}(\mathbb{R})$, which are given as follows:

$$m((a, b)) = \int_a^b g(x)\mathrm{d}x, \quad n((c, d)) = \int_c^d h(y)\mathrm{d}y;$$

then on $\mathcal{I}(\mathbb{R}^2)$ we define a measure μ by

$$\mu(D) = \int_D \int g(x)h(y)\mathrm{d}x\mathrm{d}y$$

where $D = (a, b) \times (c, d)$. Then we have

$$\mu((a, b) \times (c, d)) = \left(\int_a^b g(x)\mathrm{d}x \right) \left(\int_c^d h(y)\mathrm{d}y \right)$$

so that μ is the product measure $m \times n$.

It is important to appreciate that not all measures on $\mathcal{I}(\mathbb{R}^2)$ are product measures; for example, a very general class of measures is obtained by defining

$$\mu(D) = \int_D \int f(x, y) dx dy \qquad (9.2)$$

where $f(x, y) \geq 0$ for all $(x, y) \in \mathbb{R}^2$. Clearly, if we take $f(x, y) = \sin^2(xy)$ or $(x^2 + y^2)^{-1/2}$, then we do not obtain a product measure.

Again, the ideas in this section extend to n-fold Cartesian products. So if $A = X_{j=1}^n A_j$, then the product Boolean algebra $X_{j=1}^n \mathcal{B}(A_j)$ is the smallest Boolean algebra containing all the products $B_1 \times B_2 \times \cdots \times B_n$, where each $B_j \in A_j$. In particular, $\mathcal{I}(\mathbb{R}^n)$ is built up out of all the open hypercubes $(a_1, b_1) \times (a_2, b_2) \times \cdots \times (a_n, b_n)$, where each (a_j, b_j) is an open interval in $\mathbb{R}$ (note that these really are open cubes in the case $n = 3$).

In general, if m_j is a measure on A_j for $1 \leq j \leq n$, then the product measure $\mu = m_1 \times m_2 \times \cdots \times m_n$ on $X_{j=1}^n A_j$ is given by

$$\mu(B_1 \times B_2 \times \cdots \times B_n) = m_1(B_1) \times m_2(B_2) \times \cdots \times m_n(B_n).$$

9.3 Distributions of random vectors

In Chapter 8, we considered random variables defined on some probability space and taking values in some subset of $\mathbb{R}$. In the preceding two sections, we have done the groundwork which enables us to extend these ideas to random variables taking values in subsets of $\mathbb{R}^2$ or even $\mathbb{R}^n$. One way of approaching such random variables is to regard them as random vectors $W = X\mathbf{i} + Y\mathbf{j}$ or $W = (X, Y)$ where X and Y are random variables taking values on some subset of $\mathbb{R}$ (so X and Y are random components of the vector W). [*] In this chapter we will consider random vectors W taking values in some region D in $\mathbb{R}^2$. The law of W is then the probability measure p_W defined by

$$p_W(A) = P(W \in A) \qquad (9.3)$$

where A is in some convenient Boolean algebra of subsets of D.

Now suppose that in (9.3), $A = B \times C$ is a product. We then have

$$p_W(A) = P((X, Y) \in B \times C)$$
$$= P((X \in B) \cap (Y \in C)), \qquad (9.4)$$

that is, in this case, $p_W(A)$ is the joint probability that X is in A and Y is in B. This suggests that properties of the random vector W can also be used to investigate the relationship between the two random variables X and Y – this is a theme which we will take up again in the next section.

---

[*] $\mathbf{i}$ and $\mathbf{j}$ are the unit vectors which start from the origin and point along the x and y axes, respectively.

Note: It is common to employ the notation

$$P(X \in B, Y \in C) = P((X \in B) \cap (Y \in C))$$

and we will use this below.

For most of the examples in this chapter we will take D to be an open rectangle $(a, b) \times (c, d)$. In this case we are interested in events of the form $W \in A$, where A is in the Boolean algebra $\mathcal{I}(D) = \mathcal{I}((a, b)) \times \mathcal{I}((c, d))$. We say that W is *continuously distributed* if

$$p_W(\{p\} \times B) = p_W(A \times \{q\}) = 0$$

whenever $a \leq p \leq b, \ c \leq q \leq d, \ A \in \mathcal{I}(a, b)$ and $B \in \mathcal{I}(c, d)$ so that, in particular, if $r = (p, q)$ is a point in D, then

$$p_W(\{r\}) = 0.$$

Just as in the one-dimensional case, continuously distributed random variables take values on isolated points with zero probability so in two dimensions they take values on isolated points or lines with zero probability. In fact, if $a \leq e \leq f \leq b$ and $c \leq g < h \leq d$ we have

$$p_W([e, f] \times [g, h]) = p_W((e, f) \times (g, h)),$$

as

$$[e, f] \times [g, h] = ((e, f) \times (g, h)) \cup (\{e\} \times [g, h]) \cup (\{f\} \times [g, h])$$
$$\cup ((e, f) \times \{g\}) \cup ((e, f) \times \{h\}).$$

Given that W is distributed continuously and taking into account Exercise 9.5, we see that the law of W, p_W is determined by its values on finite unions of disjoint open rectangles contained in D.

Consider the open rectangle $(a, x) \times (c, y)$, where $x \leq b$ and $y \leq d$. The *cumulative distribution* of W is the function F_W of two variables defined by

$$F_W(x, y) = p_W((a, x) \times (c, y))$$
$$= P(X \leq x, Y \leq y). \tag{9.5}$$

Some of the properties of F_W are explored in Exercise 9.5.

We will be mainly interested in continuously distributed random vectors W which have a *probability density function (pdf)*, which is a function f_W of two variables such that for each $a \leq x \leq b, c \leq y \leq d$, we have

$$F_W(x, y) = \int_a^x \int_c^y f_W(u, v)dudv. \tag{9.6}$$

Clearly, in order to be a pdf, f_W must satisfy the conditions:

(i) $f_W(x, y) \geq 0$ for all $a \leq x \leq b, c \leq y \leq d,$

(ii) $\int_a^b \int_c^d f_W(u, v)du\,dv = 1.$

As in the one-dimensional case, we represent probabilities as areas under the curve given by the pdf, so, in the two-dimensional case, probabilities are represented as *volumes* under the surface sketched by f_W (Fig. 9.5). Just as in the one-dimensional case, f_W may be recovered from F_W by differentiation, that is

$$f_W(x, y) = \frac{\partial^2 F_W(x, y)}{\partial x \, \partial y}$$

and so f_W (if it exists) is the unique pdf associated to W.

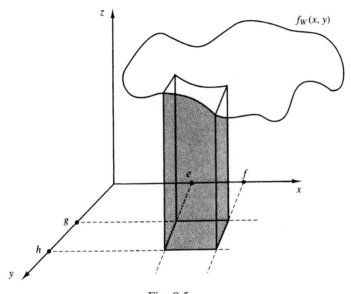

Fig. 9.5.

Example 9.3 A random vector $W = (X, Y)$ has range $(0, 1) \times (1, 2)$ and pdf

$$f_W(x, y) = \frac{1}{3}(1 + 4x^2 y).$$

Obtain the cumulative distribution of W and hence confirm that f_W is a legitimate pdf.

Solution By (9.6)

$$F_W(x, y) = \int_0^x \int_1^y \frac{1}{3}(1 + 4u^2 v)du\,dv$$

$$= \frac{1}{3}\int_0^x \left\{ \int_1^y (1 + 4u^2 v)dv \right\} du$$

$$= \frac{1}{3}x(y - 1) + \frac{2}{9}x^2(y^2 - 1)$$

then $F_W(1, 2) = 1$, as is required.

Example 9.4 [uniform random vector in D]. To begin with, let D be any region in $\mathbb{R}^2$; then we say that W is *uniformly distributed* in D if for any suitable subregion $A \subseteq D$ (Fig. 9.6), we have

$$p_W(A) = \frac{area\ of\ A}{area\ of\ D}. \tag{9.7}$$

(See Example 9.2.)

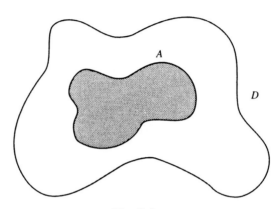

Fig. 9.6.

In the case where $D = (a, b) \times (c, d)$ (Fig. 9.7), p_W has a pdf given by

$$f_W(x, y) = \frac{1}{(b - a)(d - c)}. \tag{9.8}$$

This is the obvious generalisation of Example 8.1. You can convince yourself that (9.8) is compatible with (9.7) in Exercise 9.8.

Example 9.5 [the bivariate normal distribution] The *bivariate normal random vector* (Fig. 9.8) with parameters μ_1, $\mu_2 \in \mathbb{R}$, σ_1, $\sigma_2 > 0$ and ρ satisfying $-1 < \rho < 1$ has the pdf defined on $\mathbb{R}^2$ as

$$f_W(x, y) = \frac{1}{2\pi\sigma_1\sigma_2\sqrt{(1 - \rho^2)}} \exp\left\{ -\frac{1}{2(1 - \rho^2)} \left[\left(\frac{x - \mu_1}{\sigma_1}\right)^2 \right.\right.$$

$$\left.\left. -2\rho \left(\frac{x - \mu_1}{\sigma_1}\right) \left(\frac{y - \mu_2}{\sigma_2}\right) + \left(\frac{y - \mu_2}{\sigma_2}\right)^2 \right] \right\}. \tag{9.9}$$

In the next section, we will see that μ_1 and μ_2 can be interpreted as means, σ_1 and σ_2 as standard deviations and in Exercise 9.15 we see that ρ is a correlation coefficient. We will also confirm that (9.9) really does define a legitimate pdf.

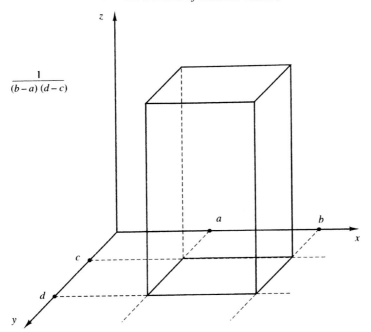

Fig. 9.7.

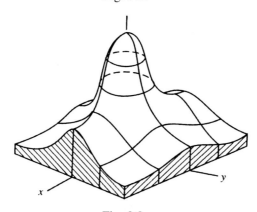

Fig. 9.8.

If we put $\mu_1 = \mu_2 = 0$ and $\sigma_1 = \sigma_2 = 1$ in (9.9), we obtain the family of *standard* bivariate normal distributions (indexed by ρ). If, on the other hand, we put $\rho = 0$ in (9.9), it should be an easy exercise to check that $f_W(x, y) = f_X(x)f_Y(y)$, where f_X is the pdf for $X \sim N(\mu_1, \sigma_1^2)$ and f_Y is the pdf for $Y \sim N(\mu_2, \sigma_2^2)$.

Note: The most general class of bivariate normals is best defined with a little linear algebra. To this end, let a, b, $c \in \mathbb{R}$ and satisfy the condition: $ac - b^2 > 0$.

Let C be the matrix $\begin{pmatrix} a & b \\ b & c \end{pmatrix}$ and note that its determinant $\det(C) = ac - b^2$ is positive. Now let w denote the column vector $\begin{pmatrix} x-\mu_1 \\ y-\mu_2 \end{pmatrix}$ and consider the quadratic form $Q(x, y) = w^\mathsf{T} C w$, where w^T is the row vector corresponding to w, so

$$Q(x, y) = a(x - \mu_1)^2 + 2b(x - \mu_1)(y - \mu_2) + c(y - \mu_2)^2;$$

then the most general bivariate normal has the pdf

$$f_W(x, y) = \frac{(\det(C))^{1/2}}{2\pi} \exp\left\{-\frac{1}{2}Q(x, y)\right\}. \tag{9.10}$$

To obtain (9.9) from (9.10) you put

$$a = \frac{1}{\sigma_1^2(1 - \rho^2)}, \quad b = -\frac{\rho}{\sigma_1\sigma_2(1 - \rho^2)} \quad \text{and} \quad c = \frac{1}{\sigma_2^2(1 - \rho^2)}$$

The results of this section are easily extended to random vectors taking values in regions of $\mathbb{R}^n$. For example, if $W = (X_1, X_2, \cdots, X_n)$ takes values in the open hypercube $(a_1, b_1) \times (a_2, b_2) \times \cdots \times (a_n, b_n)$, and if $x_j \leq b_j$ for $1 \leq j \leq n$, then the cumulative distribution is given by

$$F(x_1, x_2, \ldots, x_n) = P(W \in (a_1, x_1) \times (a_2, x_2) \times \cdots \times (a_n, x_n)),$$

and W has a pdf f_W if

$$F(x_1, x_2, \ldots, x_n) = \int_{a_1}^{x_1} \int_{a_2}^{x_2} \cdots \int_{a_n}^{x_n} f_W(u_1, u_2, \ldots, u_n) du_1 du_2 \ldots du_n$$

where f_W is positive and must integrate to 1 on the range of W.

To get the idea of how to formulate the normal distribution in higher dimensions, it is best to think in terms of extending (9.10) using $(n \times n)$ rather than (2×2) matrices (so probability theory becomes a strong motivation for learning some linear algebra!).

An example of a discrete multivariate random vector follows.

Example 9.6 [the multinomial random vector] Consider an experience which has r different outcomes (which we call type 1, type 2, etc.) occurring with probabilities $p_1, p_2, \ldots, p_r$ so that $p_1 + p_2 + \cdots + p_r = 1$. Now suppose that we have n independent repetitions of the experience and define the random variables:

X_j is the number of outcomes of type j (for $1 \leq j \leq r$) so that we have the constraint

$$X_1 + X_2 + \cdots + X_r = n \tag{#}$$

A similar argument to that of Lemma 5.12 yields

$$P(X_1 = n_1, X_2 = n_2, \ldots, X_r = n_r) = \binom{n}{n_1, n_2, \ldots, n_r} p_1^{n_1} p_2^{n_2} \cdots p_r^{n_r} \tag{9.11}$$

(see Section 2.4 for the definition of the multinomial coefficient). The law (9.11) is called the *multinomial distribution*.

Note: By (#) we see that once the values of $(r - 1)$ of the X_js are known then the rth value is determined, so, for example

$$P(X_1 = n_1, X_2 = n_2, \ldots, X_r = n_r) = P(X_1 = n_1, X_2 = n_2, \ldots, X_{r-1} = n_{r-1}).$$

Hence in this case we should regard $W = (X_1, X_2, \ldots, X_{r-1})$ as a random vector taking values in $\mathbb{N}^{r-1}$. As an example consider the case $n = 2$, where (9.11) reduces to the binomial distribution.

Example 9.7 [See Example 7.3] The genetic code of an organism consists of the four bases A, T, C and G appearing with probabilities

$$P(A) = 0.2, \quad P(T) = 0.1, \quad P(C) = 0.4, \quad P(G) = 0.3.$$

Assuming that these appear independently, find the probability that a strand of DNA composed of ten bases includes four As, one T and three Cs (so that we must have two Gs). By (9.11) this is given by

$$\binom{10}{4, 1, 3, 2,}(0.2)^4 0.1 (0.4)^3 (0.3)^2 = 0.012.$$

9.4 Marginal distributions

Let $W = (X, Y)$ be a random vector with law p_W taking values in $(a, b) \times (c, d)$. We define two probability measures p_X on $\mathcal{I}(a, b)$ and p_Y on $\mathcal{I}(c, d)$ by

$$p_X(A) = p_W(A \times (c, d))$$

and

$$p_Y(B) = p_W((a, b) \times B) \tag{9.12}$$

where $A \in \mathcal{I}(a, b)$ and $B \in \mathcal{I}(c, d)$. The fact that p_X and p_Y really are probability measures follows easily from Exercise 9.3. p_X and p_Y are called the *marginal laws* of W. To understand their significance, think of them as the laws of the components of W obtained by projecting W on to the x and y axes, respectively (Fig. 9.9), that is

$$p_X(A) = P(X \in A) \text{ and } p_Y(B) = P(Y \in B).$$

We define the *marginal cumulative distributions* F_X and F_Y by $F_X(x) = p_X((a, x))$ and $F_Y(y) = p_Y((c, y))$, where $x \leq b$ and $y \leq d$. Hence by (9.5) and (9.12)

$$F_X(x) = F_W(x, d) \text{ and } F_Y(y) = F_W(b, y).$$

We say that W has *marginal pdfs* f_X and f_Y if

$$F_X(x) = \int_a^x f_X(u)du \quad \text{and} \quad F_Y(y) = \int_c^y f_Y(v)dv$$

where f_X and f_Y satisfy the usual conditions of being positive and integrating to unity over the appropriate range.

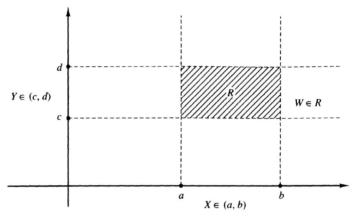

Fig. 9.9.

Lemma 9.1 *If W has a pdf f_W, then W has both of the marginal pdfs and these are given as follows*

$$f_X(x) = \int_c^d f_W(x, y)\mathrm{d}y \quad \text{and} \quad f_Y(y) = \int_a^b f_W(x, y)\mathrm{d}x.$$

Proof We will just carry this out for f_X, the argument for f_Y being similar. The marginal cumulative distribution is given by

$$F_X(x) = p_W((a, x) \times (c, d))$$
$$= \int_a^x \left[\int_c^d f_W(x, y)\mathrm{d}y \right] \mathrm{d}x$$

and the result follows.   □

Example 9.8 Return to the context of Example 9.3 and calculate:

(a) the marginal cumulative distribution functions F_X and F_Y,
(b) the marginal pdfs f_X and f_Y.

Solution

(a) Using the result of Example 9.3 we have

$$F_X(x) = F_W(x, 2) = \frac{1}{3}x(1 + 2x),$$

$$F_Y(y) = F_W(1, y) = \frac{1}{9}(y - 1)(2y + 5).$$

(b) By (8.4)

$$f_X(x) = F'_X(x) = \frac{1}{3}(1 + 4x),$$

$$f_Y(y) = F'_Y(y) = \frac{1}{9}(3 + 4y).$$

Readers should check for themselves that the relationships described in Lemma 9.1 are indeed satisfied.

Example 9.9 Find the marginal density functions f_X and f_Y for the standard bivariate normal distribution

Solution We need to calculate

$$f_Y(y) = \int_{-\infty}^{\infty} f_W(x, y)dx$$

where f_W is given by (9.9) with $\mu_1 = \mu_2 = 0$ and $\sigma_1 = \sigma_2 = 1$. The trick is to write

$$x^2 - 2\rho xy + y^2 = (x - \rho y)^2 + (1 - \rho^2)y^2,$$

so that

$$f_Y(y) = \frac{1}{\sqrt{(2\pi)}} \exp\left(-\frac{1}{2}y^2\right)$$

$$\times \frac{1}{\sqrt{[2\pi(1 - \rho^2)]}} \int_{-\infty}^{\infty} \exp\left(-\frac{1}{2}\frac{(x - \rho y)^2}{1 - \rho^2}\right) dx$$

$$= \frac{1}{\sqrt{(2\pi)}} \exp\left(-\frac{1}{2}y^2\right) \times P(-\infty \leq T \leq \infty)$$

where $T \sim N(\rho y, (1 - \rho^2))$. But $P(-\infty \leq T \leq \infty) = 1$ and so we see that f_Y is the pdf of a standard normal. The same result is easily established for f_X, that is $X \sim N(0, 1)$ and $Y \sim N(0, 1)$. If we carry through the same argument in the more general case of (9.9) a similar, although algebraically more clumsy, argument shows that $X \sim N(\mu_1, \sigma_1^2)$ and $Y \sim N(\mu_2, \sigma_2^2)$.

9.5 Independence revisited

In this section we will see how we can gain information about the relationship between a pair of random variables X and Y by thinking of them as the components of a random vector W. Specifically, we consider X with range (a, b) and Y with range (c, d) and recall from Equation (8.10) that these are probabilistically independent if

$$P(X \in A, Y \in B) = P(X \in A)P(Y \in B)$$

for all $A \in \mathcal{I}(a, b)$ and all $Y \in \mathcal{I}(c, d)$.

However, using (9.3) we can just rewrite this definition by introducing the random vector $W = (X, Y)$ taking values on $(a, b) \times (c, d)$, whose law is given as

$$p_W(A \times B) = p_X(A)p_Y(B), \qquad (9.13)$$

so, by (9.1), the measure p_W is the product of p_X and p_Y. A straightforward application of (9.12) in (9.13) establishes that p_X and p_Y (the laws of X and Y, respectively) really are the marginals of W.

Now suppose that an arbitrary random vector W has a pdf f_W, then it follows from Lemma 9.1 that each of its components X and Y have pdfs f_X and f_Y respectively.

Lemma 9.2 *X and Y are independent if and only if*

$$f_W(x, y) = f_X(x)f_Y(y) \qquad (9.14)$$

for all $(x, y) \in (a, b) \times (c, d)$.

Proof Suppose that X and Y are independent, then for all $A \times B \in \mathcal{I}(a, b) \times \mathcal{I}(c, d)$ we have, by (9.13)

$$p_W(A \times B) = \left(\int_A f_X(x)\mathrm{d}x \right) \left(\int_B f_Y(y)\mathrm{d}y \right)$$
$$= \int_{A \times B} f_X(x)f_Y(y)\mathrm{d}x\mathrm{d}y,$$

and (9.14) follows. The converse result is immediate upon integrating (9.14) over $A \times B$. $\square$

Example 9.10 Suppose that W has the bivariate normal distribution (9.8). Comparing this with the result of Example 9.9, it follows immediately from Lemma 9.2 that W has independent components if and only if $\rho = 0$.

Before proving another useful result about independence, we note that if $W = (X, Y)$ and g is any function from $\mathbb{R}^2$ to $\mathbb{R}$, then we can form the random variable $g(X, Y)$. In fact, if $\omega_1 \in S_1$, where S_1 is the sample space for X, and $\omega_2 \in S_2$, where S_2 is the sample space for Y, we have

$$g(X, Y)(\omega_1, \omega_2) = g(X(\omega_1), Y(\omega_2)).$$

We define

$$\mathbb{E}(g(X, Y)) = \int_R \int g(x, y)f_W(x, y)\mathrm{d}x\mathrm{d}y \qquad (9.15)$$

where $R = (a, b) \times (c, d)$ is the range of W.

In particular, we note that the *covariance*, $\mathrm{Cov}(X, Y)$ of X and Y is obtained by taking

$$g(X, Y) = (X - \mu_X)(Y - \mu_Y)$$

in (9.15). Properties and examples of covariance and the related concept of correlation are explored in the exercises.

Lemma 9.3 *If X and Y are independent, then*

$$\mathbb{E}(h_1(X)h_2(Y)) = \mathbb{E}(h_1(X))\mathbb{E}(h_2(Y)) \qquad (9.16)$$

for all functions h_1 defined on (a, b) and functions h_2 on (c, d).

Proof By (9.16) and (9.15), we have

$$\mathbb{E}(h_1(X)h_2(Y)) = \int_R \int h_1(x)h_2(y)f_W(x, y)\mathrm{d}x\mathrm{d}y$$

$$= \int_R \int h_1(x)h_2(y)f_X(x)f_Y(y)\mathrm{d}x\mathrm{d}y$$

$$= \left(\int_a^b h_1(x)f_X(x)\mathrm{d}x \right) \left(\int_c^d h_2(x)f_Y(y)\mathrm{d}y \right)$$

$$= \mathbb{E}(h_1(X))\mathbb{E}(h_2(Y))$$

□

Notes

(i) Of course, (9.16) is only valid for functions $h_j (j = 1, 2)$ for which both sides of (9.16) are finite.

(ii) Lemma 9.3 has a converse to the effect that if X and Y are random variables for which (9.16) holds for all h_1 and h_2, then X and Y are independent. You can try to prove this for yourself in Exercise 9.19.

We are now in a position to generalise Exercise 5.39 to the continuous case. Recall that we already used this result in our proof of the central limit theorem (Theorem 8.8). So let X and Y be independent random variables and let $M_{X+Y}(t)$ be the moment generating function of $X + Y$ (where $t \in \mathbb{R}$).

Corollary 9.4 *If X and Y are independent random variables, then for all $t \in \mathbb{R}$ we have*

$$M_{X+Y}(t) = M_X(t)M_Y(t). \qquad (9.17)$$

Proof

$$M_{X+Y}(t) = \mathbb{E}(e^{t(X+Y)})$$

$$= \mathbb{E}(e^{tX}e^{tY})$$

$$= \mathbb{E}(e^{tX})\mathbb{E}(e^{tY}) \qquad \text{by Lemma 9.3}$$

$$= M_X(t)M_Y(t).$$

□

A straightforward inductive argument now establishes the analogue of Exercise 5.39(b) in this context.

9.6 Conditional densities and conditional entropy

Let X and Y be two random variables with ranges (a, b) and (c, d) and pdfs f_X and f_Y (respectively). Suppose that X and Y are related in some way and we want to know how Y is affected if X takes one of its values x (say). Using the ideas we developed in Section 4.3, it seems that we want conditional probabilities of the form

$$P_{X=x}(Y \in B) = \frac{P((Y \in B) \cap (X = x))}{P(X = x)}$$

where $B \in \mathcal{I}(c, d)$.

However, since $P(X = x) = 0$, such conditional probabilities cannot exist and so we need some alternative way of approaching the problem. Instead of conditioning on the event $X = x$, which has zero probability, we will condition on the event $A_h = (X \in (x, x + h))$, where h is some small number, and we will see what happens in the limit as $h \to 0$.

We introduce the random vector $W = (X, Y)$ and assume that it has a joint pdf f_W on $(a, b) \times (c, d)$ so that f_X and f_Y are now interpreted as marginal pdfs. We will also make the assumption (which is always satisfied in practice) that $f_X(x) > 0$ for all $x \in (a, b)$. Hence

$$P_{A_h} = \frac{P(X \in A_h, Y \in B)}{P(X \in A_h)}$$

$$= \frac{\int_x^{x+h} \int_B f_W(u, y) du dy}{\int_x^{x+h} f_X(u) du}.$$

Now, by the fundamental theorem of calculus, we have

$$\lim_{h \to 0} P_{A_h} = \frac{\int_B f_W(x, y) dy}{f_X(x)}$$

$$= \int_B \frac{f_W(x, y)}{f_X(x)} dy.$$

Inspired by the above discussion, we define a function f_{Y_x} on (c, d) by

$$f_{Y_x}(y) = \frac{f_W(x, y)}{f_X(x)}. \tag{9.18}$$

The first thing to observe is that f_{Y_x} is a pdf on (c, d). To see this note that by Lemma 9.1 we have, for $f_X(x) \neq 0$

$$\int_c^d f_{Y_x}(y) dy = \frac{1}{f_X(x)} \int_c^d f_W(x, y) dy = \frac{1}{f_X(x)} \cdot f_X(x) = 1.$$

We now introduce the random variable Y_x whose pdf is f_{Y_x}. We call f_{Y_x} the *conditional pdf of Y given that X = x*. Note that for $B \in \mathcal{I}(c, d)$, we have

$$P(Y_x \in B) = \int_B f_{Y_x}(y)dy$$

However, it is important that this probability is not confused with $P_{X=x}(Y \in B)$, which we have already seen does not make sense. The random variable Y_x can be given a geometrical interpretation as describing observations in the $(x-y)$-plane along the line $X = x$ (Fig. 9.10).

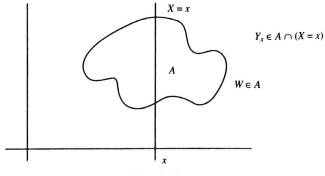

Fig. 9.10.

Notes

(i) We have seen that in the continuous case we can condition random variables, although we cannot condition probabilities on events of the form $X = x$.

(ii) Many textbooks use the notation

$$f_{Y_x}(y) = f_{Y/X}(y/x)$$

We have avoided this as it leads to confusion about which variable is being operated on by our function. However, the formula for f_{Y_x} will in general contain x as well as y and it should be understood that in this context y is the variable and x is constant.

Observe that if the random variables X and Y are independent, then via Lemma 9.2 (9.18) yields

$$f_{Y_x}(y) = f_Y(y) \quad \text{for all } y \in (c, d),$$

so that the conditional and marginal pdfs coincide and so $Y_x = Y$. You can establish a converse to this for yourself in Exercise 9.18.

Example 9.11 Find f_{Y_x} when f_W is as given in Example 9.3.

Solution Using (9.18) and the results of Examples 9.3 and 9.8, we find

$$f_{Y_x} = \frac{3x(y-1) + 2x^2(y^2-1)}{3(1+4x)}.$$

Example 9.12 Find f_{Y_x} for the bivariate normal distribution of (9.9).

Solution We have already seen in Example 9.9 that f_X is the pdf of a $N(\mu_1, \sigma_1^2)$. Hence, by (9.18), we obtain

$$f_{Y_x} = \frac{1}{\sqrt{[2\pi\sigma_2(1-\rho^2)]}} \exp\left(\frac{1}{2}\frac{[(y-\mu_2)-\frac{\sigma_1}{\sigma_2}\rho(x-\mu_1)]^2}{\sigma_2^2(1-\rho^2)}\right)$$

so that $Y_x \sim N(\nu, \omega^2)$, where $\nu = \mu_2 + \frac{\sigma_2}{\sigma_1}\rho(x-\mu_1)$ and $\omega^2 = \sigma_2^2(1-\rho^2)$.

Just as we formed the random variable Y_x by conditioning on $X = x$, so we can form a random variable X_y with range (a, b) by conditioning on $Y = y$. If we assume that $f_Y(y) > 0$ on (a, b), then X_y has pdf f_{X_y} given by

$$f_{X_y}(x) = \frac{f_W(x, y)}{f_Y(y)}. \qquad (9.19)$$

A similar argument to that of Example 9.11 above shows that, for the random vector of Example 9.3, we have

$$f_{X_y}(x) = \frac{3x(y-1) + 2x^2(y^2-1)}{3+4y}$$

and for the bivariate normal distribution of (9.9), we find that $X_y \sim N(\mu_1 + \rho(y - \mu_2), \sigma_1^2(1-\rho^2))$.

Comparing (9.19) with (9.18) yields *Bayes' formula for conditional densities* (see Theorem 4.3(a)), which is a valuable tool in statistics

$$f_{X_y}(x) = \frac{f_{Y_x}(y)f_X(x)}{f_Y(y)}. \qquad (9.20)$$

We will also find it useful to introduce a random variable X_x where we condition on the event $X = x$. Clearly, X_x is a discrete random variable with law

$$P(X_x = x) = 1, \ P(X_x = y) = 0 \quad \text{for } y \neq x.$$

We now turn our attention to entropy. We have already defined the continuous analogue of this concept in Section 8.7. Now we want to generalise Formulae (6.6) and (6.7) to obtain continuous versions of conditional entropy. We follow the same path as in the discrete case. So, given the random vector $W = (X, Y)$ with pdf f_W, we define the *conditional entropy of Y given that X = x*, which we denote as $H_x(Y)$ by

$$H_x(Y) = H(Y_x)$$

$$= -\int_c^d f_{Y_x}(y)\log(f_{Y_x}(y))dy \qquad (9.21)$$

by (8.25). We regard $H.(Y)$ as a function of the random variable X, so that $H.(Y)$ has range $\{H(Y_x),\ x \in (a, b)\}$ and pdf given by the marginal f_X. We now define the *conditional entropy of Y given X*, which we denote as $H_X(Y)$ by

$$H_X(Y) = \mathbb{E}(H.(Y)).$$

Thus we find that, by (9.18)

$$H_X(Y) = -\int_a^b f_X(x) \left(\int_c^b f_{Y_x}(y) \log(f_{Y_x}(y)) dy \right) dx$$

$$= -\int_a^b \int_c^d f_W(x, y) \log(f_{Y_x}(y)) dx dy. \qquad (9.22)$$

Note that $H_Y(X)$ is defined similarly, by interchanging the roles of X and Y in the definition.

We will study some of the properties of $H_X(Y)$ in the exercises. Here, we will content ourselves with an example.

Example 9.13 Find $H_X(Y)$ when $W = (X, Y)$ is the bivariate normal distribution.

Solution Rather than using (9.22) directly, we note that, by Example 9.12, Y_x is normal with variance $\sigma_2^2(1 - \rho^2)$. So by equation (8.26) in Example 8.11, we have

$$H_x(Y) = \log(\sigma_2(2\pi e(1 - \rho^2))^{1/2}).$$

Observe that this is a constant function (of x) and hence

$$H_X(Y) = \mathbb{E}(H.(Y)) = \log(\sigma_2(2\pi e(1 - \rho^2))^{1/2}).$$

We now give two results which we will find of use in the next section.

First of all let X, Y and Z all be random variables with range $\mathbb{R}$ and pdfs f_X, f_Y and f_Z, respectively, such that:

(i) $Y = X + Z$,
(ii) X and Z are independent.

In the next section we will consider this as a model for a communication channel wherein X is the input, Z is the noise and Y is the output.

Suppose, now, that we condition on the event $(X = x)$, then, clearly

$$Y_x = X_x + Z_x.$$

However, since X and Z are independent, $f_{Z_x} = f_Z$ and hence $Z_x = Z$.

Lemma 9.5

$$f_{Y_x}(y) = f_Z(y - x)$$

for all $y \in \mathbb{R}$.

Proof Since X and Z are independent, it follows that X_x and Z are as well; hence, for $A \in \mathcal{I}(\mathbb{R})$, we have

$$
\begin{aligned}
P(Y_x \in A) &= P(Z \in A - \{x\}, X_x = x) \\
&= P(Z \in A - \{x\}) P(X_x = x) \\
&= \int_{A-\{x\}} f_Z(y) dy \\
&= \int_A f_Z(y - x) dy \quad \text{by substitution.}
\end{aligned}
$$

But $P(Y_x \in A) = \int_A f_{Y_x}(y) \, dy$ and the result follows.   □

Corollary 9.6

$$
H_X(Y) = H(Z).
$$

Proof Apply Lemma 9.5 in (9.21) to obtain

$$
\begin{aligned}
H_X(Y) &= -\int_{\mathbb{R}} f_Z(y - x) \log(f_Z(y - x)) dy \\
&= -\int_{\mathbb{R}} f_Z(y) \log(f_Z(y)) dy \quad \text{by substitution} \\
&= H(Z).
\end{aligned}
$$

Hence, $H_X(Y) = \mathbb{E}(H.(Y)) = E(H(Z)) = H(Z).$ □

Corollary 9.6 has the nice interpretation that once the input has been specified, any uncertainty in the output is due entirely to the presence of noise.

9.7 Mutual information and channel capacity

Suppose that $W = (X, Y)$ is a random vector with range $(a, b) \times (c, d)$ and pdf f_W. To define the *mutual information* between X and Y, $I(X, Y)$, we generalise the result of Theorem 6.7(a) and define

$$
I(X, Y) = \int_a^b \int_c^d f_W(x, y) \log\left(\frac{f_W(x, y)}{f_X(x) f_Y(y)}\right) dx dy. \tag{9.23}
$$

Clearly, we have $I(X, Y) = I(Y, X)$.

We now show that the analogue of (6.9) holds.

Lemma 9.7

$$
I(X, Y) = H(Y) - H_X(Y). \tag{9.24}
$$

Proof Rewriting (9.23) yields

$$
I(X, Y) = -\int_a^b \int_c^d f_W(x, y) \log(f_Y(y)) dx \, dy
$$
$$
+ \int_a^b \int_c^d f_W(x, y) \log\left(\frac{f_W(x, y)}{f_X(x)}\right) dx \, dy
$$
$$
= -\int_c^d \left(\int_a^b f_W(x, y) dx\right) \log(f_Y(y)) dy
$$
$$
+ \int_a^b \int_c^d f_W(x, y) \log(f_{Y_x}(y)) dx \, dy \qquad \text{by (9.16)}.
$$

and the required result follows by Lemma 9.1 and (9.22). □

A similar argument to the above shows also that

$$
I(X, Y) = H(X) - H_Y(X).
$$

Mutual information is a much better behaved concept than entropy for random variables with continuous ranges, and for this reason the great mathematician A. N. Kolmogorov suggested that it be considered as the fundamental concept in information transmission. To justify this, note first of all that, unlike the entropy, we always have

$$
I(X, Y) \geq 0
$$

with equality if and only if X and Y are independent. You can verify this result for yourself in Exercise 9.28.

Further evidence comes from observing how mutual information behaves under the process of discretisation and integration. To this end let $\mathcal{P} = a \leq x_0 < x_1 < \cdots < x_n = b$ be a partition of (a, b) and let $\mathcal{Q} = c \leq y_0 < y_1 < \cdots < y_m = d$ be a partition of (c, d). We introduce, as usual, the discrete random variables $\hat{X}$ and $\hat{Y}$, with laws $\{p_1, p_2, \ldots, p_n\}$ and $\{q_1, q_2, \ldots, q_m\}$, respectively, where each

$$
p_j = \int_{x_{j-1}}^{x_j} f_X(x) dx \qquad \text{and} \qquad q_k = \int_{y_{k-1}}^{y_k} f_Y(y) dy
$$

for $1 \leq j \leq n$ and $1 \leq k \leq m$. We also introduce the joint probability distribution of $\hat{X}$ and $\hat{Y}$

$$
p_{jk} = \int_{x_{j-1}}^{x_j} \int_{y_{k-1}}^{y_k} f_W(x, y) dx dy
$$

where $1 \leq j \leq n$ and $1 \leq k \leq m$.

Now if $\Delta x_j = x_j - x_{j-1}$ and $\Delta y_k = y_k - y_{k-1}$ for $1 \le j \le n$ and $1 \le k \le m$, we find by Theorem 6.7(a) that

$$I(\hat{X}, \hat{Y}) = \sum_{j=1}^{n} \sum_{k=1}^{m} p_{jk} \log \left(\frac{p_{jk}}{p_j q_k} \right)$$

$$= \sum_{j=1}^{n} \sum_{k=1}^{m} \frac{1}{\Delta x_j \Delta y_k} p_{jk} \log \left(\frac{\frac{1}{\Delta x_j \Delta y_k} p_{jk}}{\frac{1}{\Delta x_j} p_j \cdot \frac{1}{\Delta y_k} q_k} \right) \Delta x_j \Delta y_k$$

and as we take limits (one for each variable) we find by applying the fundamental theorem of calculus that

$$\lim I(\hat{X}, \hat{Y}) = I(X, Y)$$

as is required.

Example 9.14 Suppose that W is jointly normally distributed with pdf given by (9.9). Find $I(X, Y)$.

Solution By (9.24), Example 9.13 and (8.26) we find

$$I(X, Y) = H(Y) - H_X(Y)$$
$$= \frac{1}{2} \log(2\pi e \sigma_2^2) - \frac{1}{2} \log(2\pi e \sigma_2^2 (1 - \rho^2))$$
$$= -\frac{1}{2} \log(1 - \rho^2).$$

Note that, by Exercise 9.17, we have the nice interpretation that $I(X, Y)$ depends only on the correlation coefficient ρ between X and Y. In particular, the stronger the correlation, the greater is the information transmission between X and Y.

We now consider the analogue of the set-up of Chapter 7, whereby a signal is transmitted from a source X to a receiver Y through a channel which is corrupted by the presence of noise. The difference now is that we are sending a continuous signal (such as a waveform) across the channel and so both X and Y will be random variables with continuous ranges. Imitating the idea of Section 7.2, we define the *channel capacity C* by

$$C = \max I(X, Y)$$

where the maximum is over all possible probability laws of X.

Just as was the case when we maximised the entropy in Section 8.7, we obtain different results if we apply different constraints. We will make the following natural assumptions about our input X:

$$\text{(i) } \mathbb{E}(X) = 0 \quad \text{and} \quad \text{(ii) } \mathrm{Var}(X) = \sigma_X^2.$$

(Note that even if $\mathbb{E}(X) = \mu_X$, you can always change units and replace X by $X - \mu_X$.)

We will work with a simple model of *additive noise* wherein

$$Y = X + Z$$

where both X and Z are independent. We will assume that Z is normally distributed with mean zero and variance σ_Z^2. From our work in the preceding chapter – particularly the central limit theorem – we know that this is quite a reasonable assumption to make about the noise.

Note: In many books on information theory, the variances σ_X^2 and σ_Z^2 are called the *power* of the signal and noise (respectively). This is because in many practical situations these are measured in terms of the voltages which produce them, and the power of a random voltage V with mean zero passing through a resistance R is

$$\frac{1}{R}\mathbb{E}(V^2).$$

So σ_X^2 and σ_Z^2 are genuine measures of power per unit resistance.

Theorem 9.8 *The channel capacity C is attained when X is normally distributed. Furthermore, we then have*

$$C = \frac{1}{2}\log\left(1 + \frac{\sigma_X^2}{\sigma_Z^2}\right). \tag{9.25}$$

Proof By Lemma 9.7, Corollary 9.6 and (8.26), we have

$$\begin{aligned} I(X, Y) &= H(Y) - H_X(Y) \\ &= H(Y) - H(Z) \\ &= H(Y) - \frac{1}{2}\log(2\pi e \sigma_Z^2). \end{aligned}$$

We now want

$$C = \max(I(X, Y)) = \max\left(H(Y) - \frac{1}{2}\log(2\pi e \sigma_Z^2)\right).$$

Now by Theorem 8.9, we know that the maximum value of $H(Y)$ is obtained when Y is normally distributed but $Y = X + Z$ is normal provided X is normal (see Exercise 9.22) and, furthermore, by Exercise 9.16 we have

$$\mathrm{Var}(Y) = \mathrm{Var}(X) + \mathrm{Var}(Z) = \sigma_X^2 + \sigma_Z^2.$$

Hence by (8.26) we find that $\max(H(Y)) = \frac{1}{2}\log(2\pi e(\sigma_X^2 + \sigma_Z^2))$ and

$$C = \frac{1}{2}\log(2\pi e(\sigma_X^2 + \sigma_Z^2)) - \frac{1}{2}\log(2\pi e(\sigma_Z^2)),$$

from which the result follows. $\qquad\square$

By (9.25) we see that the ability of the channel to transmit information increases in strength as the ratio of σ_X^2 to σ_Z^2, that is when the power of the incoming signal is stronger than the power of the disturbing noise.

Exercises

9.1. If A and B are finite sets, what is $\#(A \times B)$? Deduce that counting measure is a product measure.

9.2. The rigorous definition of an ordered pair is

$$(a, b) = \{\{a\}, \{a, b\}\}.$$

Show that $(a, b) = (b, a)$ if and only if $a = b$.

9.3. Convince yourself (e.g. by drawing pictures) that

$$A \times (B \cup C) = (A \times B) \cup (A \times C)$$

and

$$A \times (B \cap C) = (A \times B) \cap (A \times C).$$

9.4. Show that if A and B are finite sets, then

$$\mathcal{P}(A) \times \mathcal{P}(B) = \mathcal{P}(A \times B).$$

9.5. If $R = (a, b) \times (c, d)$ is a rectangle in $\mathbb{R}^2$, show that $\overline{R}$ can be written as the union of at most four rectangles. [*Hint*: Draw a picture.]

9.6. Let F be the cumulative distribution function of the random vector $W = (X, Y)$ on $(a, b) \times (c, d)$. Show that:

   (a) $F(x_1, y) \leq F(x_2, y)$ whenever $x_1 \leq x_2$,
   (b) $F(x, y_1) \leq F(x, y_2)$ whenever $y_1 \leq y_2$,
   (c) $F(b, d) = 1$.

9.7. A worm is placed within a rectangular box whose floor is a rectangle of dimensions $5\,\text{cm} \times 9\,\text{cm}$. Assuming a uniform distribution, find the probability that the worm is located within the area shown in Fig. 9.11.

9.8. A random variable X is uniformly distributed on the rectangle $[3, 5] \times [2, 9]$:

   (a) Write down its pdf.
   (b) Obtain the cumulative distribution function.
   (c) Calculate the probability that X takes values in the rectangle $[3, 4] \times [4, 7]$.

9.9. Find the constant C such that the following are pdfs on the given region:

   (a) Cxy^2 on $(0, 1) \times (2, 3)$,
   (b) $\frac{3}{14}(x^2 + Cxy + y^2)$ on $(1, 2)^2$.

9.10. A random vector W has the pdf on $[0, \infty)^2$ given by

$$f_W(x, y) = e^{-x-y}.$$

(a) Find the cumulative distribution.
(b) Find both marginal distributions.
(c) Are the components X and Y of W independent?

9.11. A random vector W has pdf on $[0, \infty)^2$ given by

$$f_W(x, y) = y \exp\left(-\frac{1}{2}y(4x + 1)\right).$$

(a) Check that f_W really is a pdf.
(b) Calculate both marginal pdfs.
(c) Obtain the conditional pdfs for both Y_x and X_y.

9.12. The conditional cumulative distribution function of Y given x is defined as

$$F_{Y_x}(y) = P(Y_x \leq y)$$
$$= \int_c^y f_{Y_x}(y)dy$$

if Y has a density on (c, d). Find F_{Y_x} for the example of Exercise 9.11 above.

9.13. A random vector takes all its values in the triangle bounded by the lines $x = 0$, $y = 0$ and $x + y = 1$ and the pdf of W is required to be a constant:

(a) Find the value of this constant.
(b) Calculate the joint probabilities $P(X \leq 0.5, Y \leq 0.25)$ and $P(X \leq 0.5, Y \leq 1)$.
(c) Obtain both marginal pdfs
(d) Obtain both conditional pdfs
(e) Are X and Y independent?

(*Note*: In (c) and (d) you should think carefully about how to define marginal and conditional pdfs when the region is no longer a product.)

9.14. The law of a discrete random vector $W = (X, Y)$ taking values in $\mathbb{N}^2$ is given by

$$p_W(m, n) = P(X = m, Y = n)$$
$$= \frac{\lambda^m \mu^n}{m! n!} e^{-(\lambda + \mu)}$$

where the parameters $\lambda, \mu > 0$. Find the marginal laws of X and Y and investigate whether or not these random variables are independent. Do the laws of X and Y seem familiar?

9.15. A fair die is thrown 12 times in succession. Treating these as independent trials, find the probability of obtaining each number (from 1 to 6) twice.

9.16. If the correlation coefficient between two components X and Y of a random vector W is defined (just as in the discrete case) by

$$\rho(X, Y) = \frac{\text{Cov}(X, Y)}{\sigma_X \sigma_Y}$$

show that Theorems 5.7–5.11 all extend to the case where X and Y are components of a random vector with a pdf.

9.17. Show that when X and Y are components of the bivariate normal random vector, then

$$\rho(X, Y) = \rho.$$

Hence, deduce that such random variables are independent if and only if

$$\mathbb{E}(XY) = \mathbb{E}(X)\mathbb{E}(Y).$$

9.18. Show that if $f_{Y_x} = f_Y$ for all $x \in (a, b)$, then X and Y are independent.

9.19.* (a) Prove that $\mathbb{E}(\chi_A(X)) = p_X(A)$ (where χ_A is the indicator function of A so that $\chi_A(X) = 1$ if $X \in A$ and 0 otherwise).

  (b) By writing $h_1 = \chi_A$ and $h_2 = \chi_B$, establish the converse to Lemma 9.3.

9.20. Let X and Y be independent random variables on $\mathbb{R}$ with pdfs f_X and f_Y, respectively, and let $Z = X + Y$. Show that Z has a pdf f_Z which is given by the *convolution* of f_X and f_Y, that is

$$f_Z(z) = \int_{-\infty}^{\infty} f_X(x) f_Y(z - x) \mathrm{d}x$$

(this is the continuous analogue of (by (5.8))). [*Hint*: Write the cumulative distribution $F_z(z)$ as a double integral of f_W over the region $\{(x, y) \in \mathbb{R}^2; x + y \le z\}$ and then make an appropriate substitution.]

9.21. Show that if $X \sim \mathcal{E}(1)$ and $Y \sim \mathcal{E}(1)$ are independent, then $X + Y$ has a gamma distribution with parameters $\alpha = 2$ and $\beta = 1$ (see Exercise 8.11 for the definition of the gamma distribution).

9.22. Show that if X and Y are independent and normally distributed, then $X + Y$ is also normally distributed, and find its mean and variance. [*Hint*: You can do this by calculating the convolution of the densities, but it's quicker to use Corollary 9.4.]

9.23. If X and Y are random variables, the *conditional expectation* $\mathbb{E}_x(Y)$ of Y given x is defined by

$$\mathbb{E}_x(Y) = \mathbb{E}(Y_x)$$

$$= \int_c^d y f_{Y_x}(y) \mathrm{d}y \qquad \text{(see Exercise 5.40).}$$

Find $\mathbb{E}_x(Y)$ if X and Y are components of the bivariate normal distribution.

9.24. Regard $\mathbb{E}_x(Y)$ as the value (when $X = x$) of a random variable $\mathbb{E}_X(Y)$. Note that $\mathbb{E}_X(Y)$ is a function of X and so

$$\mathbb{E}(\mathbb{E}_X(Y)) = \int_a^b \mathbb{E}_x(Y) f_X(x) dx.$$

Show that

$$\mathbb{E}(\mathbb{E}_X(Y)) = \mathbb{E}(Y).$$

9.25. If X and Y are independent, show that

$$H_X(Y) = H(Y).$$

9.26. If $W = (X, Y)$ is a random vector, define its joint entropy to be

$$H(X, Y) = -\int_a^b \int_c^d f_W(x, y) \log(f_W(x, y)) dx dy.$$

Show that

$$H(X, Y) = H(X) + H_X(Y)$$

$$= H(Y) + H_Y(X) \qquad \text{(see Theorem 6.5).}$$

(Perhaps $H(W)$ is a better notation than $H(X, Y)$?)

9.27. Use the result of Exercise 9.26 above to calculate $H(W)$ when W is the bivariate normal random vector.

9.28. (a) Show that

$$I(X, Y) = H(X) + H(Y) - H(X, Y).$$

(b) Prove that $I(X, Y) \geq 0$ with equality if and only if X and Y are independent.

9.29. A channel with input X and output Y is acted on by multiplicative noise Z where X and Z are independent, so that

$$Y = XZ.$$

(i) Show that

$$f_{Y_x}(y) = x f_Z(yx).$$

(ii) Hence deduce that

$$H_X(Y) = H(Z) - \mathbb{E}(\log(X)).$$

9.30. The input into a communication channel varies between V_1 and V_2 volts and the output takes values between W_1 and W_2 volts. If the joint probability density of the channel is

$$f_W(x, y) = \frac{1}{(V_2 - V_1)(W_2 - W_1)}$$

calculate:

(a) the input and output entropies $H(X)$ and $H(Y)$,
(b) the joint entropy $H(X, Y)$,
(c) the mutual information $I(X, Y)$.

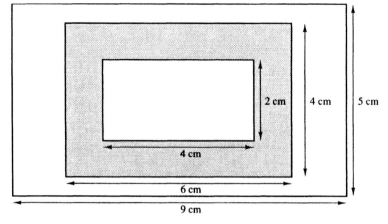

Fig. 9.11.

9.31. A communication channel is characterised by the pdfs

$$f_Y(y) = 2(1 - y) \text{ and } f_{Y_x}(y) = \frac{1}{1 - x}$$

on the triangle bounded by the lines $x = 0$, $y = 0$ and $x + y = 1$ (see Example 9.13). Calculate:

(a) $H(Y)$, (b) $H_X(Y)$, (c) $I(X, Y)$.

Further reading

All the books cited at the end of Chapter 8 are also appropriate for this chapter.

10

Markov chains and their entropy

10.1 Stochastic processes

So far in this book we have tended to deal with one (or at most two) random variables at a time. In many concrete situations, we want to study the interaction of 'chance' with 'time', e.g. the behaviour of shares in a company on the stock market, the spread of an epidemic or the movement of a pollen grain in water (Brownian motion). To model this, we need a family of random variables (all defined on the same probability space), $(X(t), t \geq 0)$, where $X(t)$ represents, for example, the value of the share at time t.

$(X(t), t \geq 0)$ is called a (continuous time) *stochastic process* or *random process*. The word 'stochastic' comes from the Greek for 'pertaining to chance'. Quite often, we will just use the word 'process' for short.

For many studies, both theoretical and practical, we discretise time and replace the continuous interval $[0, \infty)$ with the discrete set $\mathbb{Z}_+ = \mathbb{N} \cup \{0\}$ or sometimes $\mathbb{N}$. We then have a (discrete time) stochastic process $(X_n, n \in \mathbb{Z}_+)$. We will focus entirely on the discrete time case in this chapter.

Note. Be aware that $X(t)$ and X_t (and similarly $X(n)$ and X_n) are both used interchangeably in the literature on this subject.

There is no general theory of stochastic processes worth developing at this level. It is usual to focus on certain classes of process which have interesting properties for either theoretical development, practical application, or both of these. We will study Markov chains in this chapter. These are named in honour of the Russian mathematician Andrei Andreyevitch Markov (1856–1922) who first investigated them. This is a very rich class of stochastic processes, which is easily accessible to beginners and has the advantage of being theoretically interesting as well as having a wide range of applications – including information theory.

Before we begin our studies of Markov chains we will say just a little bit more about general stochastic processes. In Chapters 5 and 8 (respectively) we have met

217

random variables whose values are always non-negative integers (e.g. the binomial random variable) or always real numbers (e.g. the normal distribution). For stochastic processes, it makes sense to require that all the component random variables take values in a given set S called the *state space*. Typically in this chapter this will be a subset (or all) of $\mathbb{N}$ or $\mathbb{Z}$. For more advanced work, S may be a subset of $\mathbb{R}$ or $\mathbb{R}^n$ or be a set with some special structure such as a group.

One of the simplest examples of a discrete time stochastic process is a random walk and we have already met this process in Chapter 5. We recall that it is constructed as follows: let $\{Y_n, n \in \mathbb{N}\}$ be a set of i.i.d. random variables, each of which only takes two possible values -1 and 1, so that each

$$P(Y_n = 1) = p, \quad P(Y_n = -1) = q = 1 - p,$$

where $0 \le p \le 1$. A *random walk* is the stochastic process $(X_n, n \in \mathbb{Z}_+)$ where $X_0 = 0$ and for $n \ge 1$

$$X_n = Y_1 + Y_2 + \cdots + Y_n.$$

As we saw in Chapter 5, X_n takes values between $-n$ and n, so the state space S is $\mathbb{Z}$. In the case where $p = 1/2$, we say that we have a *symmetric* random walk.

Figure 10.1 shows two 'sample path' simulations from a symmetric random walk.

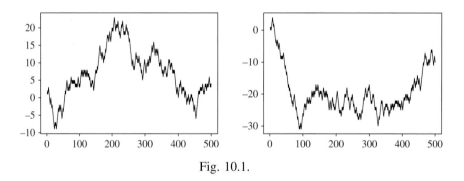

Fig. 10.1.

and Fig. 10.2 two sample path simulations from an asymmetric random walk wherein $p = 0.6$ and $q = 0.4$.

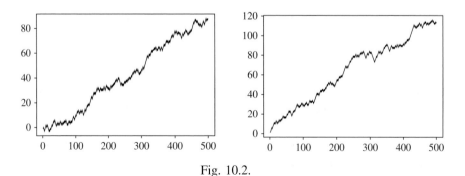

Fig. 10.2.

You can experiment further yourself by using the following *S*-plus instructions:

```
t = 500,
z = sample(c(−1, 1), size = t, replace = T, prob = c(q, p)),
x = cumsum(z),
tsplot(x).
```

Of course, you can change the sample size (500 here) to any value you like and the variables p and q *must* be replaced by numerical values.

Now let $X = (X_n, n \in \mathbb{Z}_+)$ be an arbitrary stochastic process with state space S. Let $i_0, i_1, \ldots, i_n$ be arbitrary points in S. We are often interested in probabilities of events where, for example, the process is at the point i_0 at time n_0 and it is at the point i_1 at time $n_1, \ldots$, and it is at point i_k at time n_k. This is the probability of the intersection of $k + 1$ sets and it can be written formally as

$$P((X_{n_0} = i_0) \cap (X_{n_1} = i_1) \cap \cdots \cap (X_{n_k} = i_k)).$$

This is a bit long winded and we will instead write this probability as follows (see Section 9.3)

$$P(X_{n_0} = i_0, X_{n_1} = i_1, \ldots, X_{n_k} = i_k).$$

As it is rather cumbersome when B is a complicated set involving lots of X_ns, we will also abandon our notation of $P_B(A)$ for the conditional probability of A given B and instead we will write the more conventional $P(A|B)$. So we have, for example

$$P(X_{n_k} = i_k | X_{n_0} = i_0, \ldots, X_{n_{k-1}} = i_{k-1})$$
$$= \frac{P(X_{n_0} = i_0, \ldots, X_{n_{k-1}} = i_{k-1}, X_{n_k} = i_k)}{P(X_{n_0} = i_0, \ldots, X_{n_{k-1}} = i_{k-1})}.$$

10.2 Markov chains

10.2.1 Definition and examples

In general a stochastic process has the 'Markov property' if:

In any prediction of the future, knowledge of the entire past is of the same value as knowledge of the present.

Many of the most popular stochastic processes used in both practical and theoretical work have this property. If $S \subseteq \mathbb{R}$, we say that we have a *Markov process* and if $S \subseteq \mathbb{N}$ or $\mathbb{Z}_+$, then we say that we have a *Markov chain*.

We will discretise time to be $\mathbb{Z}_+ = \{0, 1, 2, \ldots\}$, and in this chapter we'll work solely with *discrete time Markov chains*.

We'll now give a formal definition. To do this think of time instants:

$$0, 1, 2, \ldots, n-1 \text{ to be 'the past'},$$
$$n \text{ to be 'the present'},$$
$$n+1 \text{ to be 'the future'}.$$

$k_0, k_1, \ldots, k_{n+1}$ are arbitrary integers in the discrete state space S.

Definition A stochastic process $(X_n, n \in \mathbb{Z}_+)$ is called a *Markov chain* if

$$P(X_{n+1} = k_{n+1} | X_n = k_n, X_{n-1} = k_{n-1}, \ldots, X_1 = k_1, X_0 = k_0)$$
$$= P(X_{n+1} = k_{n+1} | X_n = k_n)$$

for all $k_0, k_1, \ldots, k_{n+1} \in S$ and all $n \in \mathbb{N}$.

The probability that $X_{n+1} = j$ given that $X_n = i$ is called the *(one-step) transition probability* and we write

$$P_{ij}^{n,n+1} = P(X_{n+1} = j | X_n = i).$$

In all the examples we'll consider, we will have *stationary transition probabilities*, that is,

$$P_{ij}^{n,n+1} = P_{ij} \tag{10.1}$$

is the same for all values of n, so the probability of going from i to j in one second (say) is the same starting out at time 1, time 2, time 3 etc. We will assume that (10.1) holds for the remainder of this chapter. Markov chains which have this property are sometimes called *homogeneous* in the literature to distinguish them from the general case.

Note. Sometimes it makes the notation clearer to write P_{ij} as $P_{i,j}$.

Suppose that $S = \mathbb{Z}_+$. We can collect together all the transition probabilities into a matrix $P = (P_{ij})$ called the *transition probability matrix* or sometimes *transition matrix* for short:

$$P = \begin{pmatrix} P_{00} & P_{01} & P_{02} & P_{03} & \cdots \\ P_{10} & P_{11} & P_{12} & P_{13} & \cdots \\ P_{20} & P_{21} & P_{22} & P_{23} & \cdots \\ P_{30} & P_{31} & P_{32} & P_{33} & \cdots \\ \cdot & \cdot & \cdot & \cdot & \cdot \\ \cdot & \cdot & \cdot & \cdot & \cdot \\ \cdot & \cdot & \cdot & \cdot & \cdot \\ P_{i0} & P_{i1} & P_{i2} & P_{i3} & \cdots \\ \cdot & \cdot & \cdot & \cdot & \cdot \\ \cdot & \cdot & \cdot & \cdot & \cdot \\ \cdot & \cdot & \cdot & \cdot & \cdot \end{pmatrix}$$

As we'll begin to see shortly, the matrix P plays a very important role in the theory of Markov chains. Now as S is infinite, P is an infinite matrix, which is pretty hard to deal with mathematically. Even so, we have two properties which automatically hold

(i) For all $i, j \in S$, $\quad P_{ij} \geq 0 \dots$.
(ii) For all $i \in S$, $\quad \sum_{j=1}^{\infty} P_{ij} = 1 \dots$.

Indeed (i) follows the fact that each P_{ij} is a probability, while (ii) is the natural extension of Theorem 4.2(a) to the case where the partition consists of the infinitely many disjoint events $(X_n = 0), (X_n = 1), \dots, (X_n = k), \dots$ (see also Exercise 4.10).

Any square matrix which satisfies both (i) and (ii) (i.e. it has non-negative entries and its row-sums are all unity) is said to be *stochastic* – irrespective of whether or not it is associated with a Markov chain. Note that by (i) and (ii) together, it follows that if P is a stochastic matrix, then $0 \leq P_{ij} \leq 1$ for all $i, j \in S$.

In many interesting examples, S will be finite and have say N elements. In this case P is an $N \times N$ matrix and we say that $X = (X_n, n \in \mathbb{Z}_+)$ is a *finite state Markov chain*. One of the reasons why Markov chain theory is interesting mathematically is the nice interplay between probability and matrix algebra that we get in this case. We will develop the first few steps in this. A brief survey of all the concepts in matrix algebra that are needed for this chapter can be found in Appendix 5. Before we begin developing the theory, we present some examples of Markov chains.

Example 10.1 A communication system
A binary symbol (0 or 1) is transmitted through a cascade of binary symmetric channels that are connected in series. We obtain a stochastic process $(X_n, n \in \mathbb{Z}_+)$ where X_0 is the symbol sent out and X_n is that which is received in the nth channel. The probability of successful transmission (i.e. $0 \to 0$ or $1 \to 1$) is $1 - p$, while the probability of error (i.e. $0 \to 1$ or $1 \to 0$) is p, where $0 \leq p \leq 1$ (in practice p should be small).

We can model this as a discrete time Markov chain with state space $S = \{0, 1\}$. The transition matrix is

$$P = \begin{pmatrix} P_{00} & P_{01} \\ P_{10} & P_{11} \end{pmatrix} = \begin{pmatrix} 1 - p & p \\ p & 1 - p \end{pmatrix}.$$

In this simple example, not only the rows, but also the columns sum to 1. In general, stochastic matrices which have this property are called *doubly stochastic*. We will meet them again later in this chapter but it is worth pointing out at this stage that they are the exception rather than the rule. You can get some insight into the structure of doubly stochastic matrices by reading Section (iv) in Appendix 5.

Example 10.2 Random walks revisited

A random walk is an example of a Markov chain where the state space is genuinely infinite. To make it easier to write down P, we'll impose the 'initial condition' $P_{01} = 1$ so that whenever the walker reaches 0 s(h)e always steps to the right. This means that 0 is a 'barrier' and the state space is $\mathbb{Z}_+$ rather than $\mathbb{Z}$. For $i \geq 0$, the only non-zero transition probabilities are

$$P_{i,i+1} = p, \quad P_{i,i-1} = 1 - p.$$

The transition matrix is then

$$P = \begin{pmatrix} 0 & 1 & 0 & 0 & \cdots & \cdots & \cdots \\ 1-p & 0 & p & 0 & \cdots & \cdots & \cdots \\ 0 & 1-p & 0 & p & 0 & \cdots & \cdots \\ 0 & 0 & 1-p & 0 & p & 0 & \cdots \\ \cdot & & \cdot & & \cdot & \cdot & \cdot \\ \cdot & & \cdot & & \cdot & \cdot & \cdot \\ \cdot & & \cdot & & \cdot & \cdot & \cdot \end{pmatrix}.$$

Example 10.3 A gambling model

Each time s(h)e plays, a gambler either wins £1 with probability p or loses with probability $1 - p$. S(h)e gives up either when s(h)e goes broke or attains a fortune of £M. We can model this as a Markov chain with state space $S = \{0, 1, 2, \ldots, M\}$ and transition probabilities $P_{00} = P_{MM} = 1$ and

$$P_{i,i+1} = p, \quad P_{i,i-1} = 1 - p.$$

Because of the similarities with the previous example, this is called a *finite state random walk*. The states 0 and M are called *absorbing*, as once you enter them you can never leave. When $M = 4$, the transition matrix takes the following form

$$P = \begin{pmatrix} 1 & 0 & 0 & 0 & 0 \\ 1-p & 0 & p & 0 & 0 \\ 0 & 1-p & 0 & p & 0 \\ 0 & 0 & 1-p & 0 & p \\ 0 & 0 & 0 & 0 & 1 \end{pmatrix}.$$

Example 10.4 Gene frequencies

Let $(X_1, X_2, X_3, \ldots)$ be the number of individuals in successive generations with a particular genetic trait, for example long legs, blue eyes etc. For simplicity, we'll take the population size M to be constant in time. If $X_n = i$, it may seem reasonable (as a first approximation) to argue that any member of the $(n+1)$th generation has the trait with probability i/M, independently of all the others. We can then model

$(X_n, n \in \mathbb{N})$ as a Markov chain with state space $S = \{0, 1, 2, \ldots, M\}$ and binomial transition probabilities

$$P_{ij} = \binom{M}{j} \left(\frac{i}{M}\right)^j \left(1 - \frac{i}{M}\right)^{M-j}$$

for each $0 \le i, j \le M$. It is a useful exercise to compute this matrix explicitly in the cases $M = 1, 2$ and 3.

10.2.2 Calculating joint probabilities

Here we use the notation π_j for the 'initial distribution' $P(X(0) = j)$ of a Markov chain. As the following result shows, if we know the transition matrix P and the initial probability distribution π, then we can calculate any joint probability:

Theorem 10.1 *If $(X_n, n = 0, 1, 2, \ldots)$ is a Markov chain, then*

$$P(X_0 = i_0, X_1 = i_1, X_2 = i_2, \ldots, X_n = i_n) = \pi_{i_0} P_{i_0,i_1} P_{i_1,i_2} \cdots P_{i_{n-1},i_n}.$$

Proof We'll carry out a proof by induction. For the initial step, observe that

$$P(X_0 = i_0, X_1 = i_1) = P(X_0 = i_0)P(X_1 = i_1 | X_0 = i_0)$$

$$= \pi_{i_0} P_{i_0,i_1}.$$

Now assume the result holds for some n, then by the Markov property

$$P(X_0 = i_0, X_1 = i_1, X_2 = i_2, \ldots, X_n = i_n, X_{n+1} = i_{n+1})$$
$$= P(X_{n+1} = i_{n+1} | X_0 = i_0, X_1 = i_1, X_2 = i_2, \ldots, X_n = i_n)$$
$$\times P(X_0 = i_0, X_1 = i_1, X_2 = i_2, \ldots, X_n = i_n)$$
$$= P(X_{n+1} = i_{n+1} | X_n = i_n)P(X_0 = i_0, X_1 = i_1, X_2 = i_2, \ldots, X_n = i_n)$$
$$= P_{i_n,i_{n+1}} . \pi_{i_0} P_{i_0,i_1} P_{i_1,i_2} \cdots P_{i_{n-1},i_n}.$$

hence the result holds for all $n \in \mathbb{N}$, by induction. □

Example 10.5 A Markov chain $(X_n, n = 0, 1, 2, \ldots)$ with state space $\{0, 1, 2\}$ has the transition matrix

$$P = \begin{pmatrix} 0.3 & 0.2 & 0.5 \\ 0.1 & 0.7 & 0.2 \\ 0.2 & 0.4 & 0.4 \end{pmatrix}$$

and initial distribution $\pi_0 = 0.25$, $\pi_1 = 0.5$ and $\pi_2 = 0.25$. Calculate the joint probability $P(X_0 = 0, X_1 = 2, X_2 = 1)$.

Solution Using Theorem 10.1, we see that

$$P(X_0 = 0, X_1 = 2, X_2 = 1) = \pi_0 P_{02} P_{21}$$

$$= 0.25 \times 0.5 \times 0.4 = 0.05.$$

10.3 The Chapman–Kolmogorov equations

We've already defined the one-step transition probabilities for a Markov chain. We can now extend this idea and define *n-step transition probabilities*

$$P_{ij}^{m,m+n} = P(X_{m+n} = j | X_m = i).$$

We'll make our usual stationarity assumption that these probabilities are the same for all values of m, that is

$$P_{ij}^{m,m+n} = P_{ij}^{(n)} \quad \text{where}$$

$$P_{ij}^{(n)} = P(X_n = j | X_0 = i).$$

We proceed as in the case $n = 1$ and (taking $S = \mathbb{Z}_+$) define the *n-step transition matrix* $P^{(n)}$ by

$$P^{(n)} = \begin{pmatrix} P_{00}^{(n)} & P_{01}^{(n)} & P_{02}^{(n)} & P_{03}^{(n)} & \cdots \\ P_{10}^{(n)} & P_{11}^{(n)} & P_{12}^{(n)} & P_{13}^{(n)} & \cdots \\ P_{20}^{(n)} & P_{21}^{(n)} & P_{22}^{(n)} & P_{23}^{(n)} & \cdots \\ P_{30}^{(n)} & P_{31}^{(n)} & P_{32}^{(n)} & P_{33}^{(n)} & \cdots \\ \cdot & \cdot & \cdot & \cdot & \cdot \\ \cdot & \cdot & \cdot & \cdot & \cdot \\ \cdot & \cdot & \cdot & \cdot & \cdot \\ P_{i0}^{(n)} & P_{i1}^{(n)} & P_{i2}^{(n)} & P_{i3}^{(n)} & \cdots \\ \cdot & \cdot & \cdot & \cdot & \cdot \\ \cdot & \cdot & \cdot & \cdot & \cdot \end{pmatrix}$$

In principle, each of the $P_{ij}^{(n)}$s can be computed from the joint probabilities using Theorem 10.1 but there is a simpler and more satisfying approach which we'll now develop. First we need a straightforward result about conditional probabilities.

Lemma 10.2 *If A, B and C are events, then*

$$P(A \cap B | C) = P(A | B \cap C) P(B | C)$$

Proof We assume that $P(C), P(B \cap C) \neq 0$. We then have

$$P(A \cap B | C) = \frac{P(A \cap B \cap C)}{P(C)}$$

$$= \frac{P(A \cap B \cap C)}{P(B \cap C)} \cdot \frac{P(B \cap C)}{P(C)}$$

$$= P(A | B \cap C) P(B | C).$$

$\square$

For the remainder of this chapter we'll simplify matters by assuming that we have a finite state space $S = \{1, \ldots, N\}$ (so we deal only with finite state Markov chains). The next result is of great importance. It lies at the heart of the elementary theory of Markov chains and demonstrates why the matrix algebra approach is so useful.

Theorem 10.3 (Chapman–Kolmogorov)

$$P_{ij}^{(m+n)} = \sum_{k=1}^{N} P_{ik}^{(n)} P_{kj}^{(m)} \tag{10.2}$$

so the matrix

$$P^{(n)} = P^n = P \times P \times \cdots \times P \quad (n \text{ times})$$

in the sense of **matrix multiplication**.

Notes

(i) This result extends to infinite state spaces. In that case, the sum on the right-hand side of (10.2) must be replaced by a (convergent) infinite series.

(ii) If A and B are $N \times N$ matrices, recall that their product $C = AB$ is well defined and the entries of C are given by the formula

$$C_{ij} = \sum_{k=1}^{N} A_{ik} B_{kj} \tag{10.3}$$

(see also Appendix 5).

(iii) The system (10.2) of N^2 equations is called the *Chapman–Kolmogorov equations* in honour of their discoverers.[†]

Proof of Theorem 10.3

$$P_{ij}^{(m+n)} = P(X_{m+n} = j | X_0 = i)$$

$$= \sum_{k=1}^{N} P(X_{m+n} = j, X_n = k | X_0 = i).$$

Now apply Lemma 10.2 with $A = (X_{n+m} = j)$, $B = (X_n = k)$ and $C = (X_0 = i)$ to get

$$P_{ij}^{(m+n)} = \sum_{k=1}^{N} P(X_{m+n} = j | X_n = k, X_0 = i) P(X_n = k | X_0 = i).$$

[†] Andrei Nikolaevich Kolmogorov (1903–87) was one of the leading mathematicians of the twentieth century. He was responsible for the foundation of probability on measure theory (as described in Chapter 4) and made important contributions to many areas of mathematics, including dynamical systems, topology, turbulence and algorithmic complexity.

Then use the Markov property within the first term in the sum to obtain

$$P_{ij}^{(m+n)} = \sum_{k=1}^{N} P(X_{m+n} = j | X_n = k) P(X_n = k | X_0 = i);$$

i.e. $$P_{ij}^{(m+n)} = \sum_{k=1}^{N} P_{kj}^{n,m+n} P_{ik}^{(n)}$$

$$= \sum_{k=1}^{N} P_{ik}^{(n)} P_{kj}^{(m)}.$$

Comparing this last equation with (10.3), we see that we have shown that

$$P^{(m+n)} = P^{(n)} \times P^{(m)}$$

in the sense of matrix multiplication, so that for example

$$P^{(2)} = P \times P = P^2.$$

The general result follows from this by iteration (or induction if you prefer). □

Example 10.6 For the Markov chain of Example 10.5, calculate

$$\text{(i) } P^2, \quad \text{(ii) } P(X_2 = 2 | X_0 = 1).$$

Solution

(i) P^2 $=$
$\begin{pmatrix} 0.3 & 0.2 & 0.5 \\ 0.1 & 0.7 & 0.2 \\ 0.2 & 0.4 & 0.4 \end{pmatrix} \begin{pmatrix} 0.3 & 0.2 & 0.5 \\ 0.1 & 0.7 & 0.2 \\ 0.2 & 0.4 & 0.4 \end{pmatrix}$

$=$
$\begin{pmatrix} 0.21 & 0.40 & 0.39 \\ 0.14 & 0.59 & 0.27 \\ 0.18 & 0.48 & 0.34 \end{pmatrix}.$

(ii) $P(X_2 = 2 | X_0 = 1) = P_{12}^{(2)} = 0.27.$

The final topic of this section is to apply the Chapman–Kolmogorov equations to find the law of X_n for any $n > 0$, from the transition matrix P and the initial condition π_0. We define $\pi_j^{(n)} = P(X_n = j)$ for $j = 1, \ldots, N$. The trick is to write the probabilities $\pi_1^{(n)}, \ldots, \pi_N^{(n)}$ as a row vector (i.e. a $1 \times N$ matrix) which we call $\pi^{(n)}$, so

$$\pi^{(n)} = (\pi_1^{(n)}, \pi_2^{(n)}, \ldots, \pi_N^{(n)}).$$

We treat $\pi^{(n)}$ as an 'unknown' which we'd like to find. We assume that we know the initial distribution $\pi^{(0)}$ where

$$\pi^{(0)} = (\pi_1^{(0)}, \pi_2^{(0)}, \ldots, \pi_N^{(0)}),$$

and that we also know the transition matrix P.

Theorem 10.4

$$\pi^{(n)} = \pi^{(0)} P^n. \tag{10.4}$$

Proof For each $1 \leq j \leq N$

$$\pi_j^{(n)} = P(X_n = j)$$

$$= \sum_{i=1}^{N} P(X_n = j, X_0 = i)$$

$$= \sum_{i=1}^{N} P(X_n = j | X_0 = i) P(X_0 = i)$$

$$= \sum_{i=1}^{N} \pi_i^{(0)} P_{ij}^{(n)}.$$

□

Example 10.7 Return to the Markov chain of Examples 5 and 6 and find the distribution of X_2 if we know that $P(X_0 = 0) = 0.3$ and $P(X_0 = 1) = 0.5$.

Solution $\pi^{(0)} = (0.3, 0.5, 0.2)$, hence we can use $\pi^{(2)} = \pi^{(0)} P^2$ to obtain

$$\pi^{(2)} = (0.3, 0.5, 0.2) \begin{pmatrix} 0.21 & 0.40 & 0.39 \\ 0.14 & 0.59 & 0.27 \\ 0.18 & 0.48 & 0.34 \end{pmatrix}$$

$$= (0.169, 0.511, 0.32),$$

that is, $P(X_2 = 0) = 0.169$, $P(X_2 = 1) = 0.511$, $P(X_2 = 2) = 0.32$.

There is a nice analogy between solving the matrix equation (10.4) to find the law $\pi^{(n)}$ at any time n from the initial distribution $\pi^{(0)}$ and the matrix P and solving a differential equation to find the value of a function at any later time from the initial condition which specifies the function at time 0 and the equation itself.

In this section, we've found it convenient to treat the law $\pi^{(n)}$ of X_n as a *probability vector*, that is a row vector whose entries are the probabilities that X_n takes each of its possible values. In future we'll reserve this terminology for any row vector with non-negative entries which sum to 1.

10.4 Stationary processes

Markov chains are the main stochastic processes to be considered in this chapter; however, before we probe further into their structure it will be useful to look at another interesting class, namely those that are *stationary*.

Let $X = (X_n, n \in \mathbb{Z}_+)$ be a stochastic process with state space $S = \{1, 2, \ldots, N\}$. We can completely determine the probabilities of all possible events occurring in the history of the process if we know the *finite-dimensional distributions*

$$P(X_{n_1} = i_1, X_{n_2} = i_2, \ldots, X_{n_k} = i_k).$$

Here k is any natural number, $n_1, n_2, \ldots, n_k$ are completely arbitrary times in $\mathbb{Z}_+$ (in no particular order) and $i_1, i_2, \ldots, i_k$ are arbitrary points in S. If we fix the times $n_1, n_2, \ldots, n_k$ for now, then we can regard the random variables $X_{n_1}, X_{n_2}, \ldots, X_{n_k}$ as the components of a random vector $X_{n_1, n_2, \ldots, n_k} = (X_{n_1}, X_{n_2}, \ldots, X_{n_k})$. We can form a new random vector $X_{n_1+m, n_2+m, \ldots, n_k+m}$ by shifting each of the times $n_1, n_2, \ldots, n_k$ to the later time $n_1 + m, n_2 + m, \ldots, n_k + m$. We say that the stochastic process is *stationary* [†] (or *strongly stationary*) if for all $k, m \in \mathbb{N}$ and all $n_1, n_2, \ldots n_k$, the random vectors $X_{n_1, n_2, \ldots, n_k}$ and $X_{n_1+m, n_2+m, \ldots, n_k+m}$ have the same multivariate distributions, that is

$$P(X_{n_1} = i_1, X_{n_2} = i_2, \ldots, X_{n_k} = i_k)$$
$$= P(X_{n_1+m} = i_1, X_{n_2+m} = i_2, \ldots, X_{n_k+m} = i_k)$$

for all $i_1, i_2, \ldots, i_k \in S$.

In particular, if $X = (X_n, n \in \mathbb{Z}_+)$ is a stationary process, then each X_n has the same probability law since for $m > n$ and $i \in S$ we have

$$P(X_m = i) = P(X_{n+m-n} = i) = P(X_n = i).$$

You should however be warned that this property on its own is not (in general) enough for a process to be stationary. In Exercise 10.13 you can check that the mean and standard deviation of the random variables in a stationary process are constant in time. You can also show that the 'autocovariance' $\text{Cov}(X_n, X_m)$ depends only on the time difference $|m - n|$. In general, any stochastic process which satisfies these three criteria is called *weakly stationary*. We will only be concerned with the stronger notion here, but bear in mind that weakly stationary processes have lots of important applications, for example to the theory of time series in statistics.

At this stage we'll give a very simple example of a stationary stochastic process:

Example 10.8 Let $X = (X_n, n \in \mathbb{Z}_+)$ be a stochastic process which comprises i.i.d. random variables. We'll show that it is stationary. First note that each X_n has

---

[†] Be aware that *stationary* processes have nothing to do with *stationary* transition probabilities for a Markov chain. This is two separate uses of the overworked word 'stationary'.

the same probability law p. By independence, for each $k, m \in \mathbb{N}, n_1, n_2, \ldots n_k \in \mathbb{Z}_+, i_1, i_2, \ldots, i_k \in S$

$$P(X_{n_1+m} = i_1, X_{n_2+m} = i_2, \ldots, X_{n_k+m} = i_k)$$
$$= P(X_{n_1+m} = i_1)P(X_{n_2+m} = i_2) \cdots P(X_{n_k+m} = i_k)$$
$$= p(i_1)p(i_2) \cdots p(i_k)$$
$$= P(X_{n_1} = i_1, X_{n_2} = i_2, \ldots, X_{n_k} = i_k)$$

on reversing the steps of the argument.

We'll construct more interesting examples of stationary processes by throwing the Markov property into the mix. Formally, a *stationary Markov chain* is a stochastic process which is simultaneously a Markov chain and a stationary process. To see how to construct these will be the aim of the next section.

10.5 Invariant distributions and stationary Markov chains

10.5.1 Invariant distributions

The key to constructing stationary Markov chains is finding invariant distributions. Let $X = (X_n, n \in \mathbb{Z}_+)$ be a Markov chain with transition probability matrix P. A probability vector ρ is an *invariant distribution* for X if

$$\rho = \rho P \qquad (10.5)$$

that is $\rho_j = \sum_{i=1}^{N} \rho_i P_{ij}$ for all $1 \leq j \leq N$.

It is called an invariant distribution as it remains 'invariant' (i.e. unchanged) when it is multiplied on the right by the matrix P. Other names for it commonly used in the literature are *stationary distribution* and *equilibrium distribution*. Note that if we iterate (10.5), we get $\rho = \rho P^n$ for all $n \in \mathbb{N}$.

In general, there is no reason why an invariant distribution should exist for a Markov chain and if it does exist, it may not be unique. We'll study the 'existence question' later on in this section.

Example 10.9 It is easy to check that $\rho = \left(\frac{5}{12}, \frac{7}{12}\right)$ is an invariant distribution for the Markov chain with transition probability matrix $P = \begin{pmatrix} 0.3 & 0.7 \\ 0.5 & 0.5 \end{pmatrix}$.

Example 10.10 The Markov chain we'll now describe is called the *Ehrenfest urn model* in honour of the Austrian physicist husband and wife team Paul Ehrenfest (1880–1933) and Tatiana Ehrenfest (1876–1964) who first proposed it. It is designed to describe the movement of individual gas molecules between two containers (or 'urns') which are connected by small holes. Suppose that the total number of molecules in both containers is r. We assume that the holes are sufficiently small

that only one molecule can pass through them at any given time and we also assume
that the probabilities of passage through the holes in either direction are given
by the uniform distribution. Let X_n be the number of gas molecules in container
one at time n. We assume that $(X_n, n \in \mathbb{Z}_+)$ is a Markov chain with state space
$\{0, 1, \ldots, r\}$ and from the above discussion we see that the only non-trivial one-step
probabilities are

$$P_{j,j-1} = P(X_n = j - 1 | X_{n-1} = j) = \frac{j}{r},$$

$$P_{j,j+1} = P(X_n = j + 1 | X_{n-1} = j) = \frac{r - j}{r}$$

for $1 \leq j \leq r - 1$. In Exercise 10.16, you can check for yourself that an invariant
distribution is

$$\rho_j = 2^{-r} \binom{r}{j}$$

for $0 \leq j \leq r$.

Here's the procedure for constructing stationary Markov chains from a given
Markov chain $X = (X_n, n \in \mathbb{Z}_+)$ with transition probability matrix P. Suppose
that an invariant distribution ρ exists. We construct a new Markov chain $X^{(S)} =
(X_n^{(S)}, n \in \mathbb{Z}_+)$ which has the same transition probability matrix P as X and having
initial distribution $\pi_S^{(0)} = \rho$ (the 'S' here stands for 'stationary'). It then follows
that all the $X_n^{(S)}$s have the same distribution ρ since by (10.5) $\pi_S^{(n)} = \rho P^n = \rho$. To
establish that the Markov chain is indeed stationary, it is sufficient to consider

$$P(X_0^{(S)} = i_0, X_1^{(S)} = i_1, \ldots, X_n^{(S)} = i_n)$$

$$= \pi_S^{(0)}(i_0) P_{i_0,i_1} P_{i_1,i_2} \cdots P_{i_{n-1},i_n} \qquad \text{(by theorem 10.1)}$$

$$= \pi_S^{(m)}(i_0) P_{i_0,i_1}^{(m,m+1)} P_{i_1,i_2}^{(m+1,m+2)} \cdots P_{i_{n-1},i_n}^{(m+n-1,m+n)} \qquad \text{(by homogeneity of the chain)}$$

$$= P(X_m^{(S)} = i_0, X_{m+1}^{(S)} = i_1, \ldots, X_{m+n}^{(S)} = i_n) \qquad \text{(by theorem 10.1 again)}$$

for all $m, n \in \mathbb{N}$.

There's an interesting approach to invariant distributions which uses eigenvalues
and eigenvectors. If we take transposes of both sides of equation (10.5), we obtain

$$P^{\mathrm{T}} \rho^{\mathrm{T}} = \rho^{\mathrm{T}} \qquad\qquad (10.6)$$

where ρ^{T} is just ρ written as a column vector. Now (10.5) tells us that ρ^{T} is an
eigenvector whose entries sum to 1 corresponding to the eigenvalue 1 of P^{T}. The
eigenvalues of P^{T} are precisely those of P but beware that they don't necessarily
have the same eigenvectors. Since 1 is always an eigenvalue of P (why?), we have
the following algorithmic approach to finding invariant distributions:

*Find the eigenvectors of P^{T} which correspond to the eigenvalue 1. If any of these
eigenvectors has non-negative entries, then normalise it (see Appendix 5.3) and this*

will be an invariant distribution. If 1 *is an eigenvalue of multiplicity* 1, *then this invariant distribution is unique.*

Example 10.11 Consider the Markov chain with transition probability matrix

$$P = \begin{pmatrix} 1 & 0 & 0 \\ 0 & 0.3 & 0.7 \\ 0 & 0.5 & 0.5 \end{pmatrix}.$$

Expanding the determinant in the usual way we see that $\det(P - \lambda I) = (\lambda - 1)^2(5\lambda + 1)$ and so $\lambda = 1$ appears as an eigenvalue of multiplicity 2. You can check that the two corresponding invariant distributions are $\left(0, \frac{5}{12}, \frac{7}{12}\right)$ and $(1, 0, 0)$.

Eigenvalues play an important role in the more advanced theory of Markov chains and if you are interested in this, then you should investigate the *Perron–Frobenius theorem* by consulting, for example, one of the texts cited at the end of the chapter.

10.5.2 The detailed balance condition

As an alternative to playing with eigenvalues, the search for 'detailed balance' yields a mechanism for finding invariant distributions that has a very natural physical interpretation. Suppose that $\gamma = (\gamma_1, \gamma_2, \ldots, \gamma_N)$ is the initial distribution of a Markov chain. The chain is said to be in detailed balance if the probability of starting at point i and then moving to j is the same as that of starting at j and then moving to i. More precisely, we say that the *detailed balance condition* holds for an arbitrary probability vector γ if

$$\gamma_i P_{ij} = \gamma_j P_{ji} \tag{10.7}$$

for all $1 \leq i, j \leq N$. From a physical point of view, the detailed balance condition is closely related to the idea of equilibrium (and also 'reversibility').

Theorem 10.5 *If γ satisfies the detailed balance condition for a Markov chain with transition probability matrix P, then it is an invariant distribution.*

Proof We want to show that $\gamma P = P$. Let $\beta = \gamma P$, then for all $1 \leq j \leq N$

$$\beta_j = \sum_{i=1}^{N} \gamma_i P_{ij}$$

$$= \sum_{i=1}^{N} \gamma_j P_{ji} \text{ by } (10.7)$$

$$= \gamma_j \sum_{i=1}^{N} P_{ji} = \gamma_j$$

as was required.   □

Example 10.12 We revisit the gambling model of Example 10.3. We have $P_{0,0} = P_{M,M} = 1$ and for $1 \leq i \leq M - 1$, $P_{i,i-1} = q$ and $P_{i,i+1} = p$ where $0 < p, q < 1$ and $p + q = 1$. All other one-step probabilities are 0. The detailed balance conditions yield

$$\gamma_i P_{i,i+1} = \gamma_{i+1} P_{i+1,i} \text{ for } 0 \leq i \leq N - 1,$$

that is, $\gamma_i p = \gamma_{i+1} q$.

By iteration, we obtain $\gamma_1 = \left(\frac{p}{q}\right) \gamma_0$, $\gamma_2 = \left(\frac{p}{q}\right)^2 \gamma_0, \cdots, \gamma_M = \left(\frac{p}{q}\right)^M \gamma_0$. We require

$$\gamma_0 + \gamma_1 + \cdots + \gamma_M = 1,$$

$$\text{i.e. } \gamma_0 \left(1 + \frac{p}{q} + \frac{p^2}{q^2} + \cdots + \frac{p^M}{q^M}\right) = 1.$$

Summing the geometric progression, we obtain (for $p \neq q$)

$$\gamma_0 = \frac{1 - \frac{p}{q}}{1 - \left(\frac{p}{q}\right)^{M+1}},$$

and so for $1 \leq j \leq M$

$$\gamma_j = \left(\frac{p}{q}\right)^j \left(\frac{1 - \frac{p}{q}}{1 - \left(\frac{p}{q}\right)^{M+1}}\right)$$

$$= p^j q^{M-j} \left(\frac{q - p}{q^{M+1} - p^{M+1}}\right)$$

$$= \frac{p^j q^{M-j}}{q^M + q^{M-1} p + \cdots + q p^{M-1} + p^M}.$$

In the case where $p = q = \frac{1}{2}$, you should check that

$$\gamma_0 = \gamma_1 = \cdots = \gamma_M = \frac{1}{M+1}.$$

10.5.3 Limiting distributions

In this section we'll consider the long-term behaviour of a Markov chain as $n \to \infty$. It turns out that some chains have the property that $\lim_{n \to \infty} P_{ij}^{(n)}$ exists for all $j \in S$ and is independent of i (so that as time goes on the chain 'forgets' that it started at the point i.) If this limit exists for all $1 \leq j \leq N$, we say that the Markov chain has a *limiting distribution* and we write $\pi_j^{(\infty)} = \lim_{n \to \infty} P_{ij}^{(n)}$. Our notation suggests that $\{\pi_j^{(\infty)}, 1 \leq j \leq n\}$ are being considered as probabilities and the next result tells us that this is justified.

Theorem 10.6 *If a limiting distribution exists, then:*

(a) $\pi^{(\infty)} = (\pi_1^{(\infty)}, \pi_2^{(\infty)}, \ldots, \pi_N^{(\infty)})$ is a probability vector.
(b) For each $1 \leq j \leq N$

$$\pi_j^{(\infty)} = \lim_{n \to \infty} \pi_j^{(n)}.$$

(c) $\pi^{(\infty)}$ is an invariant distribution.

Proof

(a) $\pi_j^{(\infty)} \geq 0$ for all $1 \leq j \leq N$ by properties of the limit. Now using the fact that all of the rows of a transition probability matrix sum to 1, we get

$$\sum_{j=1}^{N} \pi_j^{(\infty)} = \sum_{j=1}^{N} \lim_{n \to \infty} P_{ij}^{(n)} = \lim_{n \to \infty} \sum_{j=1}^{N} P_{ij}^{(n)} = 1.$$

(b) Using Theorem 10.4, we get for each $1 \leq j \leq N$

$$\lim_{n \to \infty} \pi_j^{(n)} = \lim_{n \to \infty} \sum_{i=1}^{N} \pi_i^{(0)} P_{ij}^{(n)}$$

$$= \sum_{i=1}^{N} \pi_i^{(0)} \lim_{n \to \infty} P_{ij}^{(n)}$$

$$= \pi_j^{(\infty)} \sum_{i=1}^{N} \pi_i^{(0)}$$

$$= \pi_j^{(\infty)}.$$

(c) We must show that $\pi^{(\infty)} P = \pi^{(\infty)}$. We define $\gamma = \pi^{(\infty)} P$, then for each $1 \leq j \leq N$

$$\gamma_j = \sum_{k=1}^{N} \pi_k^{(\infty)} P_{kj}$$

$$= \sum_{k=1}^{N} \lim_{n \to \infty} P_{ik}^{(n)} P_{kj}$$

$$= \lim_{n \to \infty} \sum_{k=1}^{N} P_{ik}^{(n)} P_{kj}$$

$$= \lim_{n \to \infty} P_{ij}^{(n+1)} \quad \text{by Theorem 10.3}$$

$$= \pi_j^{(\infty)}.$$

$\square$

Example 10.13 Let $P = \begin{pmatrix} 0.5 & 0.5 \\ 0.5 & 0.5 \end{pmatrix}$, then it is easily verified that $P^n = P$ for

all $n \in \mathbb{N}$ and so the limiting distribution exists and is simply $\pi_0^{(\infty)} = \pi_1^{(\infty)} = 0.5$.

In Example 10.13 and (more generally) Exercise 10.15, as the matrix P has such a simple form, it is a straightforward exercise to see that a limiting distribution exists and to find it. In general, this may not be the case and we need some more technical machinery. We will not go into details about this, but just mention the main concepts and results.

We say that the chain is *irreducible* if for each $i, j \in S$ there exists $m_1, m_2 > 0$ such that $P_{ij}^{(m_1)} > 0$ and $P_{ji}^{(m_2)} > 0$, that is if we start at the point i and wait long enough, there's a positive probability of going to any state j (and vice versa). The chain is *aperiodic* if there exists $r > 0$ such that $P_{ii}^{(n)} > 0$ for all $n \geq r$. We then have the following:

Theorem 10.7 *If the Markov chain is irreducible and aperiodic, then the limiting probability exists and is the unique invariant distribution.*

A proof of this result can be found in most standard books on Markov chains. In fact, if Theorem 10.7 holds, we can also give a precise description of the limiting probabilities. To describe how this works, we first introduce the first time that the chain visits the site j to be the random variable

$$T_j = \min\{n \in \mathbb{N}; \ X_n = j\}.$$

We then have

$$\pi_j^{(\infty)} = \frac{1}{\mathbb{E}(T_j|X_0 = j)}$$

for $1 \leq j \leq N$.

10.5.4 Doubly stochastic matrices and information theory revisited

In this section, we ask the question: Can we construct a stationary Markov chain $X = (X_n, n \in \mathbb{Z}_+)$ in which each X_n has maximum entropy? By Theorem 6.2 (and in the absence of constraints) this means that each X_n has a uniform distribution and so we require conditions that impose uniformity on the invariant distribution ρ. As we will now see, the key ingredient is that the transition matrix P be doubly stochastic, that is all the columns as well as the rows of the matrix sum to unity.

Theorem 10.8 *A Markov chain with transition matrix P has a uniform invariant distribution if and only if P is doubly stochastic.*

Proof If P is doubly stochastic, then $\sum_{i=1}^{N} P_{ij} = 1$ for all $1 \leq j \leq N$ and so

$$\sum_{i=1}^{N} \frac{1}{N} P_{ij} = \frac{1}{N} \sum_{i=1}^{N} P_{ij} = \frac{1}{N}.$$

Hence $\rho_i = \frac{1}{N}$ gives the required invariant distribution. Conversely, if we know that a uniform invariant distribution exists, then arguing as above we see that for each $1 \leq j \leq N$

$$\frac{1}{N} = \frac{1}{N} \sum_{i=1}^{N} P_{ij}$$

and so $\sum_{i=1}^{N} P_{ij} = 1$ as required. ☐

Now let $(X_n, n \in \mathbb{Z}_+)$ be a stationary Markov chain which has a doubly stochastic transition probability matrix P and let X_∞ be a uniformly distributed random variable. Recall from Exercise 6.14 (b) that the relative entropy between X_n and X_∞ is

$$D(X_n, X_\infty) = \log(N) - H(X_n).$$

It can be shown that $D(X_n, X_\infty)$ is a decreasing function of n and furthermore if the uniform distribution is the unique invariant distribution, then $\lim_{n \to \infty} D(X_n, X_\infty) = 0$. It then follows that $H(X_n)$ is an increasing function of n and that $\lim_{n \to \infty} H(X_n) = \log(N)$ and this resonates nicely with the second law of thermodynamics.[†]

10.6 Entropy rates for Markov chains

10.6.1 The chain rule

Let $(X_n, n \in \mathbb{Z}_+)$ be an arbitrary discrete time stochastic process. How can we understand the way in which information changes with time? One approach is to consider the entropy $H(X_n)$ as a function of time as we did at the end of the last section. However this is somewhat crude as we only take a snapshot at each time instant. A better approach may be to study the unfolding history of the process through its information content and this suggests that we focus on the joint entropy $H(X_0, X_1, \ldots, X_n)$. Using the simpler notation $p(i_0, i_1, \ldots, i_n)$ to denote

---

[†] For more details see Chapter 1 of the book by Cover and Thomas, which is referenced at the end of this chapter.

$P(X_0 = i_0, X_1 = i_1, \ldots, X_n = i_n)$ this is given by the natural generalisation of (6.5) from 2 to $n + 1$ variables, that is [†]

$$H(X_0, X_1, \ldots, X_n) = - \sum_{i_0, i_1, \ldots, i_n = 1}^{N} p(i_0, i_1, \ldots, i_n) \log(p(i_0, i_1, \ldots, i_n)).$$

We also need to say something about the multivariate generalisation of conditional entropy. Firstly, in the case of two random variables, to be consistent with the notation we've adopted for conditional probability in this chapter, we will write $H_X(Y)$ as $H(Y|X)$ and if we want to condition on n random variables $X_1, X_2, \ldots, X_n$, we'll define

$$H(Y|X_1, X_2, \ldots, X_n)$$

$$= - \sum_{j, i_1, i_2, \ldots, i_n = 1}^{N} P(Y = j, X_1 = i_1, X_2 = i_2, \ldots, X_n = i_n)$$

$$\times \log(P(Y = j|X_1 = i_1, X_2 = i_2, \ldots, X_n = i_n)).$$

Now in the case of two random variables, we saw in Theorem 6.5 that

$$H(X_0, X_1) = H(X_0) + H(X_1|X_0).$$

The generalisation of this is called the *chain rule*.

Theorem 10.9 (The chain rule)

$$H(X_0, X_1, \ldots, X_n) = H(X_0) + \sum_{i=1}^{n} H(X_i|X_0, X_1, \ldots, X_{i-1})$$

$$= H(X_0) + H(X_1|X_0) + H(X_2|X_0, X_1) + \cdots$$

$$+ H(X_n|X_0, X_1, \ldots, X_{n-1}).$$

Proof We proceed by induction on n. The case $n = 1$ involving two random variables is already established. Now assuming that the result holds for some n, we have

$$H(X_0, X_1, \ldots, X_n, X_{n+1})$$

$$= - \sum_{i_0, i_1, \ldots, i_n, i_{n+1} = 1}^{N} p(i_0, i_1, \ldots, i_{n+1}) \log(p(i_0, i_1, \ldots, i_{n+1}))$$

[†] Note that $\sum_{i_0, i_1, \ldots, i_n = 1}^{N}$ is shorthand for $\sum_{i_0 = 1}^{N} \sum_{i_1 = 1}^{N} \cdots \sum_{i_n = 1}^{N}$.

$$= - \sum_{i_0,i_1,\ldots,i_n,i_{n+1}=1}^{N} p(i_0, i_1, \ldots, i_{n+1})$$

$$\times \log(P(X_{n+1} = i_{n+1} | X_0 = i_0, X_1 = i_1, \ldots, X_n = i_n))$$

$$- \sum_{i_0,i_1,\ldots,i_n,i_{n+1}=1}^{N} p(i_0, i_1, \ldots, i_{n+1}) \log(p(i_0, i_1, \ldots, i_n)).$$

Now since

$$\sum_{i_0,i_1,\ldots,i_n,i_{n+1}=1}^{N} p(i_0, i_1, \ldots, i_{n+1}) \log(p(i_0, i_1, \ldots, i_n))$$

$$= \sum_{i_0,i_1,\ldots,i_n=1}^{N} p(i_0, i_1, \ldots, i_n) \log(p(i_0, i_1, \ldots, i_n)),$$

we deduce that

$$H(X_0, X_1, \ldots, X_n, X_{n+1})$$
$$= H(X_0, X_1, \ldots, X_n) + H(X_{n+1} | X_0, X_1, \ldots, X_n), \qquad (10.8)$$

and the result follows. □

Since conditional entropy cannot be negative, we see from (10.8) that for all $n \in \mathbb{N}$

$$H(X_0, X_1, \ldots, X_n) \geq H(X_0, X_1, \ldots, X_{n-1})$$

that is joint entropy increases with time. □

10.6.2 *Entropy rates*

We've argued that the joint entropy is a better mechanism for studying the flow of information through time then the 'marginal' entropy. A disadvantage of this approach is that we then have to deal with an infinite sequence of numbers. In some cases, we can replace the sequence by a single number and this is clearly preferable. To be precise, we define the *entropy rate* $h(X)$ of the stochastic process $X = (X_n, n \in \mathbb{Z}_+)$ by the prescription

$$h(X) = \lim_{n \to \infty} \frac{1}{n} H(X_0, X_1, \ldots, X_{n-1})$$

when this limit exists.

Example 10.14 If $(X_n, n \in \mathbb{Z}_+)$ consists of i.i.d. random variables, then $H(X_n) = H(X_0)$ for all $n \in \mathbb{N}$ and hence $H(X_0, X_1, \ldots, X_{n-1}) = nH(X_0)$ (see Exercise 10.19). It follows that $h(X)$ exists in this case and is equal to $H(X_0)$.

We'll now investigate two special cases of stochastic processes where the entropy rate always exists and where it also takes a pleasant mathematical form.

The stationary case

Theorem 10.10 *If $X = (X_n, n \in \mathbb{Z}_+)$ is a stationary process, then $h(X)$ exists and*

$$h(X) = \lim_{n \to \infty} H(X_{n-1}|X_0, X_1, \ldots, X_{n-2}). \tag{10.9}$$

Proof We must first show that the limit on the right-hand side of (10.9) exists. By similar reasoning to that in Exercise 6.10, we have

$$H(X_{n+1}|X_0, X_1, \ldots, X_n) \le H(X_{n+1}|X_1, X_2, \ldots, X_n)$$

for all $n \in \mathbb{N}$ (you can prove this for yourself in Exercise 10.19). Since the process X is stationary, we have

$$H(X_{n+1}|X_1, X_2, \ldots, X_n) = H(X_n|X_0, X_1, \ldots, X_{n-1})$$

for all $n \in \mathbb{N}$ (see Exercise 10.21). Combining these two results, we obtain

$$H(X_{n+1}|X_0, X_1, \ldots, X_n) \le H(X_n|X_0, X_1, \ldots, X_{n-1})$$

so that if we let $a_n = H(X_n|X_0, X_1, \ldots, X_{n-1})$, then the sequence $(a_n, m \in \mathbb{N})$ is monotonic decreasing. Since each $a_n \ge 0$ the sequence is also bounded below and so $\lim_{n\to\infty} a_n$ exists. Now it is a well-known fact from elementary analysis (see Exercise 10.17) that if the sequence $(a_n, n \in \mathbb{N})$ has a limit, then so does the sequence $\left(\frac{1}{n}\sum_{i=1}^{n} a_i, n \in \mathbb{N}\right)$ of arithmetic means and

$$\lim_{n \to \infty} a_n = \lim_{n \to \infty} \frac{1}{n} \sum_{i=1}^{n} a_i.$$

Using this fact, together with the chain rule (Theorem 10.9), we obtain

$$\lim_{n \to \infty} \frac{1}{n} H(X_0, X_1, \ldots, X_{n-1}) = \lim_{n \to \infty} \frac{1}{n} \sum_{i=0}^{n-1} H(X_i|X_0, X_1, \ldots, X_{i-1})$$

$$= \lim_{n \to \infty} H(X_{n-1}|X_0, X_1, \ldots, X_{n-2})$$

as was required.   ☐

The stationary Markov case

We now suppose that $(X_n, n \in \mathbb{Z}_+)$ is a stationary Markov chain with transition matrix P and initial distribution $\pi^{(0)}$, which is (of course) an invariant distribution. We know that the entropy rate exists in this case and is given by (10.9). However, we can get an even simpler expression for $h(X)$.

Theorem 10.11 *If* $(X_n, \in \mathbb{Z}_+)$ *is a stationary Markov chain with transition matrix* P *and initial distribution* $\pi^{(0)}$, *then*

$$h(X) = - \sum_{i,j=1}^{N} \pi_i^{(0)} P_{ij} \log(P_{ij}). \tag{10.10}$$

Proof By the Markov property we have

$$H(X_{n-1}|X_0, X_1, \ldots, X_{n-2}) = H(X_{n-1}|X_{n-2}).$$

(see Exercise 10.20) and by stationarity (see Exercise 10.21)

$$H(X_{n-1}|X_{n-2}) = H(X_1|X_0).$$

Hence (10.9) becomes

$$h(X) = H(X_1|X_0)$$

$$= - \sum_{i,j=1}^{N} P(X_0 = i, X_1 = j) \log(P_{ij})$$

$$= - \sum_{i,j=1}^{N} P(X_0 = i) P_{ij} \log(P_{ij})$$

$$= - \sum_{i,j=1}^{N} \pi_i^{(0)} P_{ij} \log(P_{ij}),$$

as was required. ▢

Example 10.15 We will compute the entropy rate for the most general two-state non-trivial stationary Markov chain. The transition matrix must take the form

$$P = \begin{pmatrix} 1 - \alpha & \alpha \\ \beta & 1 - \beta \end{pmatrix},$$

and we assume that $0 < \alpha, \beta < 1$. We apply the detailed balance condition with the row vector $\rho = (\rho_0, \rho_1)$ and the only information this yields is that

$$\rho_0 \alpha = \rho_1 \beta,$$

that is, $\frac{\rho_1}{\rho_0} = \frac{\alpha}{\beta}$. Since we must have $\rho_0 + \rho_1 = 1$, we easily deduce that the invariant distribution is $\rho_0 = \frac{\beta}{\alpha+\beta}$ and $\rho_1 = \frac{\alpha}{\alpha+\beta}$. Taking $\pi^{(0)} = \rho$ and substituting in

(10.10), we obtain

$$h(X) = -\frac{\beta}{\alpha + \beta}[(1 - \alpha)\log(1 - \alpha) + \alpha\log(\alpha)]$$

$$+ \frac{\alpha}{\alpha + \beta}[\beta\log(\beta) + (1 - \beta)\log(1 - \beta)]$$

$$= \frac{1}{\alpha + \beta}[\beta H_b(\alpha) + \alpha H_b(\beta)]$$

where H_b is the 'Bernoulli entropy' as given in Example 6.2.

Entropy rates have important applications to the dynamic version of Chapter 7, that is the transmission and coding of messages in time. Here we model the symbol transmitted at time n as a random variable X_n from a stationary Markov chain $X = (X_n, n \in \mathbb{Z}_+)$. To obtain good results, it is common to impose an additional constraint on the chain and insist that it is also *ergodic*. We won't give a definition of this concept here, but we observe that if it holds, then we have the pleasant consequence that time averages are well approximated by space averages in that if μ is the common value of $\mathbb{E}(X_n)$, then $\lim_{N \to \infty} \frac{1}{N}\sum_{i=0}^{N-1} X_{n-1} = \mu$ (with probability one). Of course, this is nothing but the strong law of large numbers if the X_ns are all independent. To go further into these matters is beyond the scope of the present book and interested readers are urged to consult the literature in the 'Further reading' section below.

Exercises

10.1. A Markov chain with state space $\{0, 1, 2\}$ has the transition probability matrix

$$P = \begin{pmatrix} 0.1 & 0.2 & 0.7 \\ 0.9 & 0.1 & 0 \\ 0.1 & 0.8 & 0.1 \end{pmatrix}$$

and initial distribution $\pi_0^{(0)} = 0.3$, $\pi_1^{(0)} = 0.4$ and $\pi_2^{(0)} = 0.3$. Calculate
(a) $P(X_0 = 0, X_1 = 1, X_2 = 2)$, (b) $P(X_0 = 1, X_1 = 0, X_2 = 0)$, (c) $P(X_1 = 2, X_2 = 1)$.

10.2. A Markov chain with state space $\{0, 1, 2\}$ has the transition probability matrix

$$P = \begin{pmatrix} 0.7 & 0.2 & 0.1 \\ 0 & 0.6 & 0.4 \\ 0.5 & 0 & 0.5 \end{pmatrix}.$$

Determine the conditional probabilities $P(X_2 = 1, X_3 = 1 | X_1 = 0)$ and $P(X_1 = 1, X_2 = 1 | X_0 = 0)$.

10.3. Define a stochastic process $(S_n, n \in \mathbb{Z}_+)$ by the prescription $S_n = X_1 + X_2 + \ldots + X_n$ where the X_ns are mutually independent. Prove that $(S_n, n \in \mathbb{Z}_+)$ is a Markov chain.

10.4. Let $(S_n, n \in \mathbb{Z}_+)$ be a symmetric random walk. Deduce that $P(S_n = a) = P(S_n = -a)$ for all $a, n \in \mathbb{Z}_+$

10.5. Let $(S_n, n \in \mathbb{Z}_+)$ be a random walk and for each $a \in \mathbb{Z}$, define T_a to be the number of steps taken for the walk to reach the point a. Clearly T_a is a random variable (it is called the 'first passage time to the point a'). If the random walk is symmetric, calculate

(a) $P(T_1 = 1)$,
(b) $P(T_2 = 2)$,
(c) $P(T_2 = 3)$,
(d) $P(T_2 = 4 | T_1 = 1)$,
(e) $P(T_2 = 4, T_1 = 1)$.

10.6. A binary message (a 0 or a 1) is sent across a channel consisting of several stages where transmission through each stage has a probability α of error. Suppose that $X_0 = 0$ is the signal that is sent and let X_n be the signal received at the nth stage. Suppose that $(X_n, n \in \mathbb{Z}_+)$ is a Markov chain with transition probabilities $P_{00} = P_{11} = 1 - \alpha$ and $P_{01} = P_{10} = \alpha$.

(a) Determine the probability that no error occurs up to stage 2.
(b) Find the probability that a correct signal is received at stage 2.

10.7. A Markov chain with state space $\{0, 1, 2\}$ has the transition probability matrix

$$P = \begin{pmatrix} 0.1 & 0.2 & 0.7 \\ 0.2 & 0.2 & 0.6 \\ 0.6 & 0.1 & 0.3 \end{pmatrix}.$$

(a) Calculate the matrix P^2, (b) Compute $P(X_3 = 1 | X_1 = 0)$, (c) Calculate $P(X_3 = 1 | X_0 = 0)$.

10.8. A Markov chain with state space $\{0, 1, 2\}$ has the transition probability matrix

$$P = \begin{pmatrix} 0.3 & 0.2 & 0.5 \\ 0.5 & 0.1 & 0.4 \\ 0.5 & 0.2 & 0.3 \end{pmatrix}$$

and initial distribution determined by $\pi_0^{(0)} = \pi_1^{(0)} = 0.5$. Find $\pi_0^{(2)}$ and $\pi_0^{(3)}$.

10.9. A Markov chain $(X_n, n \in \mathbb{Z}_+)$ with state space $\{0, 1, 2\}$ has transition probability matrix P where

$$P = \begin{pmatrix} 0.4 & 0.3 & 0.3 \\ 0.7 & 0.1 & 0.2 \\ 0.5 & 0.2 & 0.3 \end{pmatrix}$$

and initial distribution $\pi_0^{(0)} = 0.1$, $\pi_1^{(0)} = 0.3$ and $\pi_2^{(0)} = 0.6$

(a) Calculate $P(X_0 = 0, X_1 = 2, X_2 = 1)$.
(b) Obtain P^2 and P^3 and use these results to find:

(i) $P(X_3 = 2 | X_1 = 1)$,
(ii) $P(X_5 = 1 | X_2 = 0)$,
(iii) the distribution of X_2.

10.10. Return to the format of Exercise 10.6 above and compute $P(X_5 = 0|$ $X_0 = 0)$.

10.11. A particle moves on a circle through points which have been labelled 0, 1, 2, 3, 4 (in a clockwise order). At each step it has a probability p of moving to the right (clockwise) and $1 - p$ of moving to the left (anticlockwise). Let X_n denote its location on the circle after the nth step.

(a) Find the transition probability matrix of the Markov chain $(X_n, n = 0, 1, 2, \ldots)$.

(b) If the initial probability distribution is uniform and $p = 0.3$, calculate

$$P(X_0 = 1, X_1 = 2, X_2 = 3, X_3 = 4, X_4 = 0).$$

10.12. The transition probability matrix of a two-state Markov chain is given by

$$P = \begin{pmatrix} p & 1-p \\ 1-p & p \end{pmatrix}.$$

Show by induction that

$$P^n = \begin{pmatrix} \frac{1}{2} + \frac{1}{2}(2p-1)^n & \frac{1}{2} - \frac{1}{2}(2p-1)^n \\ \frac{1}{2} - \frac{1}{2}(2p-1)^n & \frac{1}{2} + \frac{1}{2}(2p-1)^n \end{pmatrix}.$$

If the state space is $\{0, 1\}$ and $p = 0.1$, find $P(X(12) = 1|X(0) = 0)$.

10.13. Let $(X_n, n \in \mathbb{Z}_+)$ be a stationary process taking values in a finite state space S. Show that:

(a) $\mathbb{E}(X_n)$ is a constant function of n,
(b) $\mathrm{Var}(X_n)$ is a constant function of n,
(c) $\mathrm{Cov}(X_n, X_m)$ depends only on $|m - n|$.

10.14. Determine which of the Markov chains with the given transition probability matrix has an invariant distribution and find it explicitly when it exists:

(a) $\begin{pmatrix} 0.7 & 0.3 \\ 0.4 & 0.6 \end{pmatrix}$,

(b) $\begin{pmatrix} \frac{1}{2} & \frac{1}{4} & \frac{1}{4} \\ \frac{1}{8} & \frac{1}{2} & \frac{3}{8} \\ \frac{3}{8} & \frac{1}{4} & \frac{3}{8} \end{pmatrix}$,

(c) $\begin{pmatrix} 0 & 1 & 0 \\ 0 & \frac{2}{3} & \frac{1}{3} \\ \frac{1}{6} & \frac{5}{6} & 0 \end{pmatrix}$.

10.15. Use the result of Exercise 10.12 to explicitly find the limiting distribution (when it exists) for the Markov chain with transition probability matrix

$$P = \begin{pmatrix} p & 1-p \\ 1-p & p \end{pmatrix}, \quad \text{where } 0 \le p \le 1.$$

10.16. Deduce that $\rho_j = 2^{-r}\binom{r}{j}$ for $0 \le j \le r$ is the invariant distribution for the Ehrenfest urn model.

10.17. Deduce that if $(a_n, n \in \mathbb{N})$ has a limit a, then $\left(\frac{1}{n}\sum_{i=1}^{n} a_i, n \in \mathbb{N}\right)$ converges to the same limit. [Hint: Show that

$$\left| a - \frac{1}{n}\sum_{i=1}^{n} a_i \right| \le \frac{1}{n}\sum_{i=1}^{N_0} |a_i - a| + \frac{1}{n}\sum_{i=N_0+1}^{n} |a_j - a|,$$

and choose N_0 to be sufficiently large as to ensure that both $\frac{1}{n}$ and $|a_n - a|$ are as small as you like when $n > N_0$.]

10.18. Let $(X_n, n \in \mathbb{Z}_+)$ be a sequence of i.i.d. random variables. Show that $H(X_0, X_1, \ldots, X_{n-1}) = nH(X_0)$.

10.19. If $X = (X_n, n \in \mathbb{Z}_+)$ is an arbitrary stochastic process, show that $H(X_{n+1}|X_0, X_1, \ldots, X_n) \le H(X_{n+1}|X_1, X_2, \ldots, X_n)$.

10.20. If $(X_n, n \in \mathbb{Z}_+)$ is a Markov chain, deduce that $H(X_n|X_0, X_1, \ldots, X_{n-1}) = H(X_n|X_{n-1})$.

10.21. If $(X_n, n \in \mathbb{Z}_+)$ is a stationary process, show that for all $n \in \mathbb{N}$: (a) $H(X_{n+1}|X_1, X_2, \ldots, X_n) = H(X_n|X_0, X_1, \ldots, X_{n-1})$, (b) $H(X_n|X_{n-1}) = H(X_1|X_0)$.

10.22. If $(X_n, , n \in \mathbb{Z}_+)$ is a stationary process, show that for all $n \in \mathbb{N}$:

(a) $H(X_n|X_0, X_1, \ldots, X_{n-1}) \le \frac{1}{n+1} H(X_0, X_1, \ldots, X_n)$.

(b) $\frac{1}{n+1} H(X_0, X_1, \ldots, X_n) \le \frac{1}{n} H(X_0, X_1, \ldots, X_{n-1})$

10.23. Calculate the entropy rates for all the stationary Markov chains associated with the invariant distributions obtained in (10.14).

10.24. If $h(X)$ is the entropy rate for a stationary Markov chain defined on a state space S which has N elements, show that

$$0 \le h(X) \le \log(N).$$

Under what conditions is equality attained here?

Further reading

There are many books that give introductory accounts of stochastic processes and Markov chains at a level that is compatible with that of this chapter. A very accessible applications-based approach can be found in Chapter 4 of S. Ross *Introduction to Probability Models* (eighth edition), Academic Press (1972, 2003). For a thorough and dedicated treatment of Markov chains, see J. Norris *Markov Chains*, Cambridge University Press (1997). In particular, here you will find a detailed proof of Theorem 10.7, which was omitted from the account given here. You will also find a good account of Markov chains in some general books on probability theory such as G. Grimmett and D. Stirzaker *Probability and Random Processes* (third edition), Oxford University Press (1982, 2001) and R. Durrett *Probability: Theory and Examples*, Duxbury Press, Wadsworth Inc (1991).

Two more specialised books are well worth looking at: F. P. Kelly *Reversibility and Stochastic Networks*, J.Wiley and Sons Inc. (1979) gives a nice account of the relationship between detailed balance and reversibility (see also Section 1.9 of Norris). O. Häggström *Finite Markov Chains and Algorithmic Applications*, London Mathematical Society Student Texts No. 52, Cambridge University Press (2002) gives a very readable account of Monte Carlo Markov chains, which have recently become important in Bayesian statistics (a brief introduction to this topic may be found in Section 5.5 of Norris.)

The key reference for entropy rates of stochastic processes is T. M. Cover, J. A. Thomas *Elements of Information Theory*, J. Wiley and Sons Inc. (1991). Some coverage of this topic is also in the book by Blahut cited at the end of Chapter 7. You might also consult Chapter 3 of *Information Theory* by J. C. A. van der Lubbe, Cambridge University Press (1997).

Exploring further

After having reached the end of this book, you should have obtained a basic toolkit to help you explore deeper. I've already given you a number of leads in the 'Further reading' sections at the end of each chapter and you should follow up those that appeal most to you. I cannot recommend too highly the magnificent *Feller*, Volume 1 for strengthening your background and after that there is always Volume 2 for more advanced topics (but note that it is *much* more advanced).

To broaden and deepen your knowledge of probability, it is essential to learn more about stochastic processes. You will find some material on this in Feller's books but if you want more modern and systematic treatments, the following are all very readable and should be quite accessible: *An Introduction to Stochastic Modelling* by H. M. Taylor and S. Karlin (Academic Press, 1994), *Introduction to Probability Models* by S. Ross (Academic Press, 1985), *Stochastic Processes* by S. Ross (J. Wiley and Sons, 1983), *Probability and Random Processes*, by G. R. Grimmett and D. R. Stirzaker (Clarendon Press, Oxford, 1992) and (at a slightly more sophisticated level), *An Introduction to Probabilistic Modelling* by P. Brémaud (Springer-Verlag, 1988) and *Adventures in Stochastic Processes* by S. I. Resnick (Birkhaüser, 1992).

All the above references tend to emphasise the more applied side of stochastic processes. If theoretical aspects are of greater interest to you, you should first make sure you are well-grounded in the measure theoretic and analytical aspects of probability theory which you can obtain by studying Itô's book, for example, or you might look at J. S. Rosenthal, *A First Look at Rigorous Probability Theory* (World Scientific, 2000). Two fascinating and complementary introductions to the modern theory of 'noise', both containing interesting applications and written at a reasonably elementary level are *Probability with Martingales* by D. Williams (Cambridge University Press, 1991) and *Stochastic Differential Equations* by B. Oksendal (Springer-Verlag, 1985). If you are interested in 'taming' the noise in the stock exchange (see below), you might also consult T. Mikosch, *Elementary Stochastic*

Calculus with Finance in View (World Scientific, 1998). David Williams, *Weighing the Odds* (Cambridge University Press, 2001) gives a highly readable and quite unique introduction to probability and statistics with some coverage of stochastic processes and a closing chapter on quantum probability and quantum computing.

Nowadays many universities are teaching courses on mathematical finance at undergraduate level. This is a subject in which remarkable advances have been made in the last 30 years, particularly in understanding option pricing and related phenomena. Probability theory plays an absolutely vital role in this – particularly the stochastic processes that are known as 'martingales' and also Brownian motion. One of the best introductions to this subject that keeps technical sophistication to a minimum is M. Baxter and A. Rennie, *Financial Calculus* (Cambridge University Press, 1996). Other accessible books in this area are M. Capiński and T. Zastawniak, *Mathematics for Finance* (Springer, 2003) and S. Neftci, *An Introduction to the Mathematics of Financial Derivatives* (Academic Press, 2000).

There are far fewer books in existence dealing with information theory than there are on probability, and I've already mentioned all of the accessible ones with which I'm familiar. Again, knowledge of stochastic processes allows you to study the flow of information over time. You can learn a lot about this from Ihara's book. *Informed Assessments* by A. Jessop (Ellis Horwood, 1995) gives a very readable undergraduate-level account of an information theoretic approach to elementary statistical inference (relying heavily, of course, on the maximum entropy principle). *Elements of Information Theory* by T. M. Cover and J. A. Thomas (J. Wiley and Sons, 1991) gives a fine introduction to a number of more advanced topics, including models of the stock market. Finally, for a fascinating (and non-technical) journey through probability, information, economics, fluid flow, the origins of chaos (and much else besides) read *Chance and Chaos* by David Ruelle (Penguin, 1991).

Appendix 1
Proof by mathematical induction

Suppose that $P(n)$ is a proposition that makes sense for every natural number $n = 1, 2, 3, \ldots$, for example we could have that $P(n)$ is either of the assertions that:

(a) $\sum_{k=1}^{n} k = \frac{1}{2}n(n+1)$,

(b) $\dfrac{d^n}{dx^n}[f(x) \cdot g(x)] = \sum_{r=0}^{n} \binom{n}{r} \dfrac{d^r}{dx^r}(f(x)) \cdot \dfrac{d^{n-r}}{dx^{n-r}}(g(x))$.

If we wanted to prove that (i) and (ii) hold, we might start trying to do it for the cases $n = 1$, $n = 2$, $n = 3$, etc., but this is very laborious and is clearly never going to deliver the required result for all n. The method of mathematical induction finds a way around this, which is encoded in the following:

Principle of mathematical induction

Suppose we can establish the following:
 (i) that $P(a)$ is valid for some natural number $a \geq 1$,
 (ii) that whenever $P(n)$ is valid then $P(n+1)$ is also valid,

then if (i) and (ii) both hold, $P(n)$ is valid for all $n \geq a$.

This principle is established as an axiom of the mathematical structure of arithmetic. We can see the sense of it as follows: once (i) establishes the validity of $P(a)$, (ii) allows us to deduce $P(a+1)$, then (ii) again yields $P(a+2)$ and we continue *ad infinitum*. In many applications we take $a = 1$ as in the example given below.

Example Prove (a) above by mathematical induction. We note that when $n = 1$, the left-hand side is 1 and the right-hand side is $\frac{1}{2}(1)(1+1) = 1$. Hence, $P(1)$ is valid.
Now assume $P(n)$; then to prove $P(n+1)$ we have

$$\sum_{k=1}^{n+1} k = \sum_{k=1}^{n} k + (n+1)$$

$$= \frac{1}{2}n(n+1) + (n+1)$$

$$= (n+1)\left(\frac{1}{2}n+1\right)$$

$$= \frac{1}{2}(n+1)(n+2)$$

and so the required result is established.

As an exercise to test your understanding, you should try to prove (b) above (usually called the Leibniz rule). To help you along the way, note that $P(1)$ is the usual rule for differentiation of a product (which you should assume). You will also find the result of Exercise 2.7 helpful here.

Appendix 2
Lagrange multipliers

We begin with a function f defined on $\mathbb{R}^3$ and suppose that we want to find its stationary values, which might be maxima, minima or saddle points. It is known that these can be obtained as the solutions of the three simultaneous equations

$$\frac{\partial f}{\partial x} = 0, \frac{\partial f}{\partial y} = 0 \quad \text{and} \quad \frac{\partial f}{\partial z} = 0.$$

Now suppose that we have a *constraint* on our function f of the form

$$g(x, y, z) = 0.$$

As an example of a practical situation where this makes sense, suppose that $f(x, y, z) = 8xyz$ is the volume of a rectangular box of dimension $2x$ times $2y$ times $2z$, which is centred at the origin. If there are no constraints we don't need calculus to deduce that the minimum possible volume is 0 and the maximum is infinite. On the other hand, suppose we ask the question: What is the maximum possible volume that can be enclosed in a sphere of radius 1 centred on the origin (Fig. A2.1)?

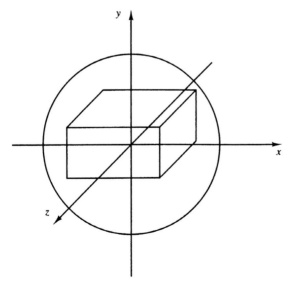

Fig. A2.1.

249

As the diagram shows, there are clearly some finite non-zero values of x, y and z for which this volume can be attained. First note that since the equation of the sphere is

$$x^2 + y^2 + z^2 = 1$$

then the constraint $g(x, y, z) = x^2 + y^2 + z^2 - 1$.

The method of Lagrange multipliers tells us that, in situations like this, we should form the function of four variables L defined by

$$L(x, y, z; \lambda) = f(x, y, z) + \lambda g(x, y, z)$$

where $\lambda \in \mathbb{R}$ is our *Lagrange multiplier*.

Our required stationary points of f, subject to the constraint g, are then the solutions of the four simultaneous equations

$$\frac{\partial L}{\partial x} = 0, \quad \frac{\partial L}{\partial y} = 0, \quad \frac{\partial L}{\partial z} = 0, \quad \frac{\partial L}{\partial \lambda} = 0.$$

Note that the final equation $\frac{\partial L}{\partial \lambda} = 0$ is just the constraint equation $g(x, y, z) = 0$.

As an application, we now solve the problem described above concerning the maximum volume of the box contained within the sphere. Here we have

$$L(x, y, z; \lambda) = 8xyz + \lambda(x^2 + y^2 + z^2 - 1)$$

and our four equations are

$$\frac{\partial L}{\partial x} = 8yz + 2\lambda x = 0, \tag{i}$$

$$\frac{\partial L}{\partial y} = 8xz + 2\lambda y = 0, \tag{ii}$$

$$\frac{\partial L}{\partial z} = 8xy + 2\lambda z = 0, \tag{iii}$$

$$\frac{\partial L}{\partial \lambda} = x^2 + y^2 + z^2 - 1 = 0. \tag{iv}$$

We substitute for λ in (i), (ii) and (iii) to obtain

$$-\frac{4yz}{x} = -\frac{4xz}{y} = -\frac{4xy}{z}.$$

Hence, $x^2 = y^2 = z^2$. Now substitute into (iv) to find

$$3x^2 = 1$$

so that $x = y = z = 3^{-1/2}$ and so the maximum volume of the box is $8 \times 3^{-3/2} = 1.54$ (to two decimal places).

We make two observations:

(i) The technique used above of eliminating λ first from the equations is typically the best approach to problems of this type. Note that λ is not itself evaluated as we don't need to know it explicitly (see Section 6.4, where the context is somewhat different).
(ii) We didn't need to apply a second test in order to check that the stationary point that we'd found was indeed a maximum as this was obvious from the context. Again this is typical of such problems.

In the above, we have formulated the method of Lagrange multipliers solely for functions in three dimensions subject to a single constraint. For the application in Section 6.4, we need a more general formalism, so let f be a function on $\mathbb{R}^n$ which is subject to the m constraints $g_i(x_1, x_2, \ldots, x_n) = 0$ for $1 \le i \le m$, where $m \le n$. Now consider the function L on $\mathbb{R}^{n+m}$ defined by

$$L(x_1, x_2, \ldots, x_n; \lambda_1, \lambda_2, \ldots, \lambda_m) = f(x_1, x_2, \ldots, x_n)$$
$$+ \sum_{j=1}^{m} \lambda_j g_j(x_1, x_2, \ldots, x_n)$$

where $\lambda_1, \lambda_2, \ldots, \lambda_m$ are the Lagrange multipliers. Then the stationary points of f subject to the m constraints can be found by solving the $(m+n)$ simultaneous equations

$$\frac{\partial L}{\partial x_j} = 0 \,(1 \le j \le n) \qquad \text{and} \qquad \frac{\partial L}{\partial \lambda_k} = 0 \,(1 \le k \le m)$$

Appendix 3
Integration of $\exp\left(-\frac{1}{2}x^2\right)$

The main point of this appendix is to justify our assertion that (8.14) really does define a pdf of a random variable. The following result employs in its proof one of my favourite tricks in the whole of mathematics.

Theorem A3.1

$$\int_{-\infty}^{\infty} \exp\left(-\frac{1}{2}x^2\right) dx = \sqrt{(2\pi)}.$$

Proof

$$\text{Write } I = \int_{-\infty}^{\infty} \exp\left(-\frac{1}{2}x^2\right) dx.$$

So

$$I^2 = \left(\int_{-\infty}^{\infty} \exp\left(-\frac{1}{2}x^2\right) dx\right) \left(\int_{-\infty}^{\infty} \exp\left(-\frac{1}{2}y^2\right) dy\right)$$

$$= \int_{-\infty}^{\infty} \int_{-\infty}^{\infty} \exp\left(-\frac{1}{2}(x^2 + y^2)\right) dxdy.$$

Now introduce polar co-ordinates $x = r\cos(\theta)$ and $y = r\sin(\theta)$ and note that the Jacobean determinant of this transformation is equal to r; hence

$$I^2 = \int_0^{2\pi} \int_0^{\infty} e^{-r^2/2} r\, dr\, d\theta = 2\pi \int_0^{\infty} e^{-r^2/2} r\, dr.$$

Upon making the substitution $u = \frac{r^2}{2}$, we obtain

$$\int_0^{\infty} e^{-r^2/2} r\, dr = \int_0^{\infty} e^{-u} du = 1$$

and the required result follows. $\qquad \square$

Corollary A3.2 *Define for $\sigma > 0$ and $\mu \in \mathbb{R}$*

$$f(x) = \frac{1}{\sigma\sqrt{(2\pi)}} \exp\left(-\frac{1}{2}\left(\frac{x-\mu}{\sigma}\right)^2\right)$$

for each $x \in \mathbb{R}$; then f is a probability density function.

252

Proof Clearly, $f(x) \geq 0$ for each $x \in \mathbb{R}$. To show that $\int_{-\infty}^{\infty} f(x) = 1$, make the standardising substitution $z = \frac{x-\mu}{\sigma}$ and then use the result of Theorem A3.1. □

Finally, we establish a result about the gamma function mentioned at the end of Chapter 2.

Corollary A3.3

$$\Gamma(1/2) = \sqrt{\pi}.$$

Proof On substituting $y = \sqrt{(2x)}$ and using Theorem A3.1, we find

$$\Gamma(1/2) = \int_0^\infty e^{-x} x^{-1/2} dx$$

$$= \sqrt{2} \int_0^\infty \exp\left(-\frac{y^2}{2}\right) dy$$

$$= \sqrt{2} \times \frac{1}{2} \times \sqrt{(2\pi)} = \sqrt{\pi}.$$

□

Appendix 4

Table of probabilities associated with the standard normal distribution

How to use this table to calculate $P(Z > z)$, where $z > 0$
  Suppose that to three significant figures $z = 1.38$. You should:

(A) Go down the far left column to find the number 1.3.
(B) Go along the top row to find the number .08.
(C) Find the unique number where the row to the right of 1.3 meets the column directly under .08. This is the number you need.

Tables of $G(z) = P(Z \geqslant z)$ for $z \geqslant 0$

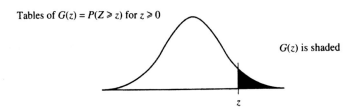

$G(z)$ is shaded

Fig. A4.1.

$$.08$$
$$*$$
$$*$$
$$*$$
$$1.3 * \quad .0838$$

So to three decimal places of accuracy

$$P(Z > 1.38) = 0.084.$$

z	.00	.01	.02	.03	.04	.05	.06	.07	.08	.09
0.0	.5000	.4960	.4920	.4880	.4840	.4801	.4761	.4721	.4681	.4641
0.1	.4602	.4562	.4522	.4483	.4443	.4404	.4364	.4325	.4286	.4247
0.2	.4207	.4168	.4129	.4090	.4052	.4013	.3974	.3936	.3897	.3859
0.3	.3821	.3783	.3745	.3707	.3669	.3632	.3594	.3557	.3520	.3483
0.4	.3446	.3409	.3372	.3336	.3300	.3264	.3228	.3192	.3156	.3121
0.5	.3085	.3050	.3015	.2981	.2946	.2912	.2877	.2843	.2810	.2776
0.6	.2743	.2709	.2676	.2643	.2611	.2578	.2546	.2514	.2483	.2451
0.7	.2420	.2389	.2358	.2327	.2296	.2266	.2236	.2206	.2177	.2148
0.8	.2119	.2090	.2061	.2033	.2005	.1977	.1949	.1921	.1894	.1867
0.9	.1841	.1814	.1788	.1762	.1736	.1711	.1685	.1660	.1635	.1611
1.0	.1587	.1562	.1539	.1515	.1492	.1469	.1446	.1423	.1401	.1379
1.1	.1357	.1335	.1314	.1292	.1271	.1251	.1230	.1210	.1190	.1170
1.2	.1151	.1131	.1112	.1093	.1075	.1056	.1038	.1020	.1003	.0985
1.3	.0968	.0951	.0934	.0918	.0901	.0885	.0869	.0853	.0838	.0823
1.4	.0808	.0793	.0778	.0764	.0749	.0735	.0721	.0708	.0694	.0681
1.5	.0668	.0655	.0643	.0630	.0618	.0606	.0594	.0582	.0571	.0559
1.6	.0548	.0537	.0526	.0516	.0505	.0495	.0485	.0475	.0465	.0455
1.7	.0446	.0436	.0427	.0418	.0409	.0401	.0392	.0384	.0375	.0367
1.8	.0359	.0351	.0344	.0366	.0329	.0322	.0314	.0307	.0301	.0294
1.9	.0287	.0281	.0274	.0268	.0262	.0256	.0250	.0244	.0239	.0233
2.0	.02275	.02222	.02169	.02118	.02067	.02018	.01970	.01923	.01876	.01831
2.1	.01786	.01743	.01700	.01659	.01618	.01578	.01539	.01500	.01463	.01426
2.2	.01390	.01355	.01321	.01287	.01255	.01222	.01191	.01160	.01130	.01101
2.3	.01072	.01044	.01017	.00990	.00964	.00939	.00914	.00889	.00866	.00842
2.4	.00820	.00798	.00776	.00755	.00734	.00714	.00695	.00676	.00657	.00639
2.5	.00621	.00604	.00587	.00570	.00554	.00539	.00523	.00508	.00494	.00480
2.6	.00466	.00453	.00440	.00427	.00415	.00402	.00391	.00379	.00368	.00357
2.7	.00347	.00336	.00326	.00317	.00307	.00298	.00289	.00280	.00272	.00264
2.8	.00255	.00248	.00240	.00233	.00226	.00219	.00212	.00205	.00199	.00193
2.9	.00187	.00181	.00175	.00169	.00164	.00159	.00154	.00149	.00144	.00139
3.0	.00135	.00131	.00126	.00122	.00118	.00114	.00111	.00107	.00103	.00100
3.1	.00097									
3.2	.00069									
3.3	.00048									
3.4	.00034									
3.5	.00023									
3.6	.00016									
3.7	.00011									
3.8	.00007									
3.9	.00005									
4.0	.00003									

Appendix 5

A rapid review of matrix algebra

This appendix gives a very concise account of all the key concepts of matrix theory that we need in Chapter 10. It is designed to be a quick reminder of facts that are already known (except for the last part) and is no substitute for systematic study of the subject.

(i) *Basic concepts*

Let $m, n \in \mathbb{N}$. A real-valued $m \times n$ matrix[†] is an array of mn real numbers that are arranged as follows

$$A = \begin{pmatrix} a_{11} & a_{12} & \cdots & a_{1n} \\ a_{21} & a_{22} & \cdots & a_{2n} \\ \cdot & \cdot & \cdots & \cdot \\ \cdot & \cdot & \cdots & \cdot \\ \cdot & \cdot & \cdots & \cdot \\ a_{m1} & a_{m2} & \cdots & a_{mn} \end{pmatrix}.$$

This matrix has m rows and n columns. For each $1 \le i \le m$, the ith row comprises the numbers $a_{i1}, a_{i2}, \ldots, a_{in}$ while the jth column consists of $a_{1j}, a_{2j}, \ldots, a_{mj}$ for each $1 \le j \le n$. The generic entry of the matrix is the number a_{ij} which lies at the intersection of the ith row and the jth column and a commonly used notation for matrices is $A = (a_{ij})$. The set of all $m \times n$ real matrices have a natural vector space structure so if B is another $m \times n$ matrix with the same generic form as A, we define addition by

$$A + B = \begin{pmatrix} a_{11} + b_{11} & a_{12} + b_{12} & \cdots & a_{1n} + b_{1n} \\ a_{21} + b_{21} & a_{22} + b_{22} & \cdots & a_{2n} + b_{2n} \\ \cdot & \cdot & \cdots & \cdot \\ \cdot & \cdot & \cdots & \cdot \\ \cdot & \cdot & \cdots & \cdot \\ a_{m1} + b_{m1} & a_{m2} + b_{m2} & \cdots & a_{mn} + b_{mn} \end{pmatrix},$$

[†] More generally one can consider complex matrices whose entries may be complex numbers or even matrices with entries in an arbitary field, but the real valued case is sufficient for our needs.

and if $\lambda \in \mathbb{R}$, scalar multiplication is defined as

$$
\lambda A = \begin{pmatrix}
\lambda a_{11} & \lambda a_{12} & \cdots & \lambda a_{1n} \\
\lambda a_{21} & \lambda a_{22} & \cdots & \lambda a_{2n} \\
\cdot & \cdot & \cdots & \cdot \\
\cdot & \cdot & \cdots & \cdot \\
\cdot & \cdot & \cdots & \cdot \\
\lambda a_{m1} & \lambda a_{m2} & \cdots & \lambda a_{mn}
\end{pmatrix}.
$$

The zero matrix O has each $a_{ij} = 0$ and so $\lambda O = O$ for all $\lambda \in \mathbb{R}$ and $A + O = O + A = A$ for each $m \times n$ matrix A.

A *row vector* v is a $1 \times n$ matrix and is usually written $v = (v_1\ v_2\ \cdots\ v_n)$ or $v = (v_1, v_2, \cdots, v_n)$. A *column vector* is an $m \times 1$ matrix and it is usually written

$$
w = \begin{pmatrix}
w_1 \\
w_2 \\
\cdot \\
\cdot \\
w_m
\end{pmatrix}.
$$

The transpose A^{T} of an $m \times n$ matrix is the $n \times m$ matrix whose generic entry is a_{ji}, so

$$
A^T = \begin{pmatrix}
a_{11} & a_{21} & \cdots & a_{m1} \\
a_{12} & a_{22} & \cdots & a_{m2} \\
\cdot & \cdot & \cdots & \cdot \\
\cdot & \cdot & \cdots & \cdot \\
a_{1n} & a_{2n} & \cdots & a_{mn}
\end{pmatrix}.
$$

Note that the transpose of a row vector is a column vector and vice versa. We have the useful identities

$$(A + \lambda B)^{\mathrm{T}} = A^{\mathrm{T}} + \lambda B^{\mathrm{T}} \text{ and } (A^{\mathrm{T}})^{\mathrm{T}} = A.$$

Matrix multiplications makes sense for the product of an $m \times n$ matrix A and a $p \times r$ matrix B if and only if $n = p$. In this case, we write $C = A \times B$ (or $C = AB$), where C is the $m \times r$ matrix whose generic entry is

$$c_{ij} = \sum_{k=1}^{n} a_{ik}b_{kj}.$$

Matrix multiplication is associative but not commutative, for example take $A = \begin{pmatrix} 0 & 1 \\ 0 & 0 \end{pmatrix}$ and $B = \begin{pmatrix} 0 & 0 \\ 1 & 0 \end{pmatrix}$ and check that $AB - BA = \begin{pmatrix} 1 & 0 \\ 0 & -1 \end{pmatrix}$. The effect of taking the transpose on matrix multiplication is summed up in the identity $(AB)^{\mathrm{T}} = B^{\mathrm{T}}A^{\mathrm{T}}$.

If A is an $m \times n$ matrix and v is a $n \times 1$ column vector, then $y = Av$ is an $m \times 1$ column vector. It is easy to see that if v_1 and v_2 are two such vectors and $\alpha_1, \alpha_2 \in \mathbb{R}$, then

$$A(\alpha_1 v_1 + \alpha_2 v_2) = \alpha_1 A v_1 + \alpha_2 A v_2$$

so that A acts as a linear mapping from the Euclidean space $\mathbb{R}^n$ into $\mathbb{R}^m$. In fact any linear mapping from $\mathbb{R}^n$ into $\mathbb{R}^m$ can be described in terms of a matrix action and this correspondence is one of the reasons why matrices are so important in mathematics.

A *square matrix* is an $n \times n$ matrix so a square matrix has equal numbers of rows and columns. Let M_n be the set of all real $n \times n$ matrices. Then $C = AB$ is well defined for any A, $B \in M_n$ and $C \in M_n$ so M_n is closed under matrix multiplication.* In particular, we write $A^2 = A \times A$ and more generally $A^n = A \times A \times \cdots \times A$ (n times). A square matrix is said to be diagonal if $a_{ij} = 0$ whenever $i \neq j$ and the identity matrix I is the diagonal matrix for which each $a_{ii} = 1$. We then have $AI = IA = A$ for all $A \in M_n$.

(ii) *Determinants*

The *determinant* of a 2×2 matrix $A = \begin{pmatrix} a_{11} & a_{12} \\ a_{21} & a_{22} \end{pmatrix}$ is the number $\det(A)$ or $|A|$ defined by

$$\det(A) = \begin{vmatrix} a_{11} & a_{12} \\ a_{21} & a_{22} \end{vmatrix} = a_{11}a_{22} - a_{12}a_{21}.$$

For a 3×3 matrix, we define its determinant by

$$\det(A) = \begin{vmatrix} a_{11} & a_{12} & a_{13} \\ a_{21} & a_{22} & a_{23} \\ a_{31} & a_{32} & a_{33} \end{vmatrix}$$

$$= a_{11} \begin{vmatrix} a_{22} & a_{23} \\ a_{32} & a_{33} \end{vmatrix} - a_{12} \begin{vmatrix} a_{21} & a_{23} \\ a_{31} & a_{33} \end{vmatrix} + a_{13} \begin{vmatrix} a_{21} & a_{22} \\ a_{31} & a_{32} \end{vmatrix}.$$

The determinant of an $n \times n$ matrix is defined recursively by the formula

$$\det(A) = \sum_{j=1}^{n} (-1)^{j+1} a_{1j} \det(A_j)$$

where A_j is the $(n-1) \times (n-1)$ matrix obtained by deleting the first row and jth column of A. Our account of determinants appears to have privileged the first row but you can in fact expand a determinant by using any row or column. Note that

$$\det(A^{\mathrm{T}}) = \det(A) \quad \text{and} \quad \det(AB) = \det(A)\det(B).$$

(iii) *Eigenvalues and eigenvectors*

Let A be a fixed $n \times n$ matrix and λ be an unknown complex number. The quantity $\det(A - \lambda I)$ is an nth degree polynomial in the unknown λ, for example if $A = \begin{pmatrix} a_{11} & a_{12} \\ a_{21} & a_{22} \end{pmatrix}$ then you can easily check that

$$\det(A - \lambda I) = \lambda^2 - (a_{11} + a_{22})\lambda + \det(A).$$

In the general case, the fundamental theorem of algebra tells us that the equation $\det(A - \lambda I) = 0$ has n complex number solutions and these numbers are called the *eigenvalues* of A. We will denote them as $\lambda_1, \lambda_2, \ldots, \lambda_n$. There is no reason why these numbers should be distinct (although they will be sometimes) and if $\lambda_1 = \lambda_2 = \cdots = \lambda_r$ (say), we say that the common value is an eigenvalue with *multiplicity r*. We

---

* As well as being a vector space, M_n is also a ring and in fact an algebra.

emphasise that even though A is a real-valued matrix, eigenvalues may be complex and observe that the complex conjugate $\bar{\lambda}$ must be an eigenvalue of A whenever λ is. A sufficient condition for A to have all of its eigenvalues being real numbers is that it is *symmetric*, that is $A = A^T$, i.e. $a_{ij} = a_{ji}$ for each $1 \leq i, j \leq n$.

In general, A^T has the same eigenvalues as A. To see this observe that $0 = \det(A - \lambda I)$ if and only if $0 = \det((A - \lambda I)^T) = \det(A^T - \lambda I)$.

It can be shown that λ is an eigenvalue of A if and only if there exists a $n \times 1$ column vector $v \neq 0$ such that $(A - \lambda I)v = 0$, that is $Av = \lambda v$ and this is the usual textbook definition of an eigenvalue. The vector v which appears in this equation is called the *eigenvector* of A corresponding to the eigenvalue λ. Note that if $c \in \mathbb{R}$ with $c \neq 0$, then cv is an eigenvector of A whenever v is – indeed, this follows easily from the linearity of A since $A(cv) = c(Av) = c(\lambda v) = \lambda(cv)$. If v is an eigenvector of A whose entries v_i are all non-negative, then we can take $c = \left(\sum_{i=1}^{n} v_i\right)^{-1}$ to obtain an eigenvector that is also a probability vector (i.e. all its entries are non-negative and these all sum to one.) Such an eigenvector will be said to be *normalised*.

(iv) *Permutation matrices*

A *permutation matrix* in M_n is one for which each row and each column has one of its entries as 1 and all others as 0. So there are only two permutation matrices in M_2, these being $\begin{pmatrix} 1 & 0 \\ 0 & 1 \end{pmatrix}$ and $\begin{pmatrix} 0 & 1 \\ 1 & 0 \end{pmatrix}$. You should write down the six permutation matrices in M_3. Note that there are $n!$ permutation matrices in M_n. The reason for the name 'permutation matrices' is that the action of such a matrix on a column vector v simply produces another column vector whose entries are a permutation of those of v, for example $\begin{pmatrix} 1 & 0 \\ 0 & 1 \end{pmatrix}\begin{pmatrix} v_1 \\ v_2 \end{pmatrix} = \begin{pmatrix} v_1 \\ v_2 \end{pmatrix}$ and $\begin{pmatrix} 0 & 1 \\ 1 & 0 \end{pmatrix}\begin{pmatrix} v_1 \\ v_2 \end{pmatrix} = \begin{pmatrix} v_2 \\ v_1 \end{pmatrix}$. *

By a *convex combination* of k matrices $A_1, A_2, \ldots, A_k \in M_n$ we mean a matrix $A \in M_n$, which has the form

$$A = \lambda_1 A_1 + \lambda_2 A_2 + \cdots + \lambda_k A_k$$

where $\lambda_j \geq 0$ $(1 \leq j \leq k)$ and $\sum_{j=1}^{k} \lambda_j = 1$. A beautiful theorem which is attributed to G. D. Birkhoff and J. von Neumann states that:

> *Every doubly stochastic matrix in M_n is a convex combination of permutation matrices.*

If we apply the Birkhoff–von Neumann theorem in the case $n = 2$, we find that the most general doubly stochastic matrix in M_2 is of the form

$$\alpha \begin{pmatrix} 1 & 0 \\ 0 & 1 \end{pmatrix} + (1 - \alpha) \begin{pmatrix} 0 & 1 \\ 1 & 0 \end{pmatrix} = \begin{pmatrix} \alpha & 1 - \alpha \\ 1 - \alpha & \alpha \end{pmatrix}$$

where $0 \leq \alpha \leq 1$. Of course this can also be easily verified directly.

---

* The permutation matrices in M_n are a representation of the symmetric group on n letters.

Selected solutions

Chapter 2

1. 12.
2. (a) 5040, (b) 720.
5. 2 598 960.
6. 12.
8. 2730.
9. 86 737.
10. 3060.
11. (a) 120, (b) 64, (c) 84.
15. 27 720.

Chapter 3

1. (a) S_1, (b) $\{1\}$, (c) $\{3,4,6\}$, (d) R_3, (e) ϕ, (f) $\{1,3,4,6\}$, (g) $\{2,3,4,5,6\}$.
2. (a) [1,4], (b) [0.5,1], (c) $[-1, 1]$.
8. $J_1 \cup J_2 = (-\infty, -9] \cup [-5, 4] \cup [6, 15]$.
11. 0.125.
13. (a) 0.25, (b) $e^{-2} = 0.135\ldots$, (c) 3, (d) 1.61.
16. Since $\mathcal{P}_A$ is a partition of A we have

$$A = (A \cap E_1) \cup (A \cap E_2) \cup \ldots \cup (A \cap E_n)$$

and the result now follows immediately from Exercises 3.9(iii).

Chapter 4

3. (c) $\frac{3}{4}$, $\frac{1}{4}$, $\frac{1}{2}$, $\frac{1}{2}$.
4. $\frac{1}{6}$.
5. 0.86.
6. (a) 0.147, (b) 0.853.

7. (a) 0.119, (b) 0.179, (c) 0.032, (d) 0.328.

13. (i) 0.318, (ii) 0.265, (iii) 0.530.

14. 0.13.

15. $Q(0) = Q(1) = \frac{1}{2}$ so that there is also maximum uncertainty of the output.

$$Q_0(0) = Q_1(1) = 1 - \varepsilon, \quad Q_0(1) = Q_1(0) = \varepsilon$$

So the $Q_i(j)$s are the same as the $P_i(j)$s and the channel is completely symmetric between input and output.

16. $Q(1) = p(1 - \varepsilon), \quad Q(E) = \varepsilon, \quad Q(0) = (1 - p)(1 - \varepsilon),$
$Q_0(0) = Q_1(1) = 1, \quad Q_E(0) = 1 - p, \quad Q_E(1) = p.$

17. $Q(1) = (1 - p)\varepsilon + p(1 - \varepsilon - \rho) = \varepsilon + p - 2p\varepsilon - p\rho,$
$Q(E) = \rho,$
$Q(0) = p\varepsilon + (1 - p)(1 - \varepsilon - \rho) = 1 - \varepsilon - p + 2p\varepsilon + p\rho - \rho,$
$Q_0(0) = \frac{(1-\varepsilon-\rho)(1-p)}{Q(0)}, \quad Q_0(1) = \frac{\varepsilon p}{Q(0)},$
$Q_E(0) = 1 - p, \quad Q_E(1) = p,$
$Q_1(1) = \frac{(1-\varepsilon-\rho)p}{Q(1)}, \quad Q_1(0) = \frac{\varepsilon(1-p)}{Q(1)}.$

18. (a) 0.366, (b) 0.0093, (c) 0.468.

19. $P(A \cap B) = P(\{4\}) = 0.25, \quad P(A)P(B) = (0.5)(0.5) = 0.25.$

20. $P(A) = P(B) = P(C) = \frac{1}{2}, \quad P(A \cap B) = P(B \cap C) = P(A \cap C) = \frac{1}{4}.$
$P(A \cap B \cap C) = 0.$

22. $Q(0) = 0.542, \ Q(1) = 0.458.$

23. $P(A) = \frac{1}{4}, \ P(B) = \frac{1}{8}, \ P(C) = \frac{1}{2}.$ These are not coherent (they sum to $\frac{7}{8}$). To make them coherent change C's odds to 5 to 3 on.

25. Solve $(\frac{5}{6})^n = \frac{1}{2}$ to find at least four throws required.

26. 0.109.

27. (a) 0.659, (b) 0.341, (c) 0.299.

28. (a) $\frac{1}{13983816}$, (b) $\frac{1}{54201}$, (c) $\frac{1}{233064}$, (d) $\frac{1}{1032}$, (e) $\frac{1}{57}$, (f) $\frac{1}{54}$, (g) 0.436.

29. $P_{T_p}(D_p) = 0.098.$ Hence our previous intuition was misguided.

Chapter 5

1. $p(0) = 0, \ p(1) = \frac{1}{9}, \ p(2) = \frac{1}{18}, \ p(3) = \frac{1}{6}, \ p(4) = \frac{1}{6}, \ p(5) = \frac{1}{2}.$

2. $p(0) = \frac{1}{8}, \ p(1) = \frac{3}{8}, \ p(2) = \frac{3}{8}, \ p(3) = \frac{1}{8}.$

4. $p(0) = 0.1591, \ p(1) = 0.4773, \ p(2) = 0.3182, \ p(3) = 0.0455$ (hypergeometric distribution).

5. $\mathbb{E}(X) = 3.889, \ \text{Var}(X) = 1.876, \ \sigma(X) = 1.370.$

18. Both equal $-0.473.$

19. (a) $\frac{3}{8}$, (b) $\frac{3}{8}$.

21. $P(S(n) = 2) = 0.335.$

22. $p_{-2,4} = p_{2,4} = p_{-1,1} = p_{1,1} = \frac{1}{4}$

    $\mathrm{Cov}(X, Y) = 0$ since $\mathbb{E}(XY) = 0$, $\mathbb{E}(X) = 0$ and $\mathbb{E}(Y) = 2.5$.

23. (a) 0.17, (b) 0.4845, (c) 0.2969, (d) 0.5155, (e) 0.8336.
24. $p = 0.65$, $n = 16$, $p(7) = 0.0442$.
27. (i) 0.3477, (ii) 0.0058, (iii) 0.5026.
30. (a) 0.24, (b) 0.0864, (c) 0.0346, (d) 0.1728.
33. (i) 0.04, (ii) 0.0017, (iii) 0.000 018.
37. Follows from the fact that

$$\frac{d^n}{dt^n}\left(e^{tx_j}\right)\Bigg|_{t=0} = x_j^n.$$

38. (b) $\mathbb{E}(X^3) = \lambda^3 + 3\lambda^2 + \lambda$, $\mathbb{E}(X^4) = \lambda^4 + 6\lambda^3 + 7\lambda^2 + \lambda$.
39. (a)

$$M_{X+Y}(t) = \sum_{j=1}^{n}\sum_{k=1}^{m} e^{(x_j+y_k)t} p_{jk}$$

$$= \sum_{j=1}^{n} e^{x_j t} p_j \sum_{k=1}^{m} e^{y_k t} q_k = M_X(t) M_Y(t).$$

(b) follows from (a) by induction.
40. (ii) Use (4.1) and Lemma 5.4(a).

Chapter 6

1. 3.32 bits, 1.74 bits and 0.74 bits.
2. (a) 5.17 bits, (b) 2.58 bits.
3. Let X be the number of 1s in the word, then $X \sim b(5, 0.6)$:

   (a) $P(X \geq 3) = 0.6826$ $I = 0.55$ bits,
   (b) $P(X \leq 4) = 0.9222$ $I = 0.117$ bits,
   (c) $P(X = 3) = 0.3456$ $I = 1.53$ bits.

4. For $x \geq 1$, $\int_1^x \frac{1}{t} dt \geq \int_1^x dt$ and for $0 \leq x < 1$ $\int_x^1 \frac{1}{t} dt < \int_x^1 dt$.
5. (a) 1.5, (b) 1.34.
8. (a) 1.78 bits, (b) 0.86 bits, (c) 0.88 bits, (d) 0.90 bits, (e) 0.02 bits.
10. By Gibbs' inequality, for each $1 \leq j \leq n$

$$H_j(Y) = -\sum_{k=1}^{m} p_j(k)\log(p_j(k)) \leq -\sum_{k=1}^{m} p_j(k)\log(q(k))$$

and the result follows by (6.7). The result for $I(X, Y)$ follows from (6.9).

12. $p_j = \frac{1}{n-1}(1 - p_1)$ for $2 \le j \le n$ thus the probabilities $p_2, p_3, \ldots, p_n$ are as 'uniformly distributed' as possible, as we would expect.

13. (a) $p_1 = 0.154$, $p_2 = 0.292$, $p_3 = 0.554$, (b) $T = 1.13 \times 10^{23}$ K.
    Is 'temperature' meaningful for such a small number of particles?

14. (a) Use Gibbs' inequality (Exercise 6.9), (c) Immediate from Theorem 6.7(a).

15. Use the result of Exercise 6.14(b).

Chapter 7

1. $I(0, 0) = -\ln(1 - p)$, $I(0, E) = I(1, E) = 0$, $I(1, 1) = -\ln(p)$.
    So $I(S, R) = (1 - \varepsilon)H(S)$ and $C = 1 - \varepsilon$, which is realised when $p = \frac{1}{2}$.

3. (a) $H(R) = -y \log(y) - (1 - y - \rho) \log(1 - y - \rho) - \rho \log(\rho)$ where
    $y = \varepsilon + p - 2p\varepsilon - p\rho$.
    (b) $H_S(R) = -(1 - \varepsilon - \rho) \log(1 - \varepsilon - \rho) - \rho \log(\rho) - \varepsilon \log(\varepsilon)$.
    (c) $I(S, R) = -y \log(y) - (1 - y - \rho) \log(1 - y - \rho) + (1 - \varepsilon - \rho) \log(1 - \varepsilon - \rho) + \varepsilon \log(\varepsilon)$.

4. (a) Differentiation shows maximum attained where $y = \frac{1}{2}(1 - \rho)$, so that $p = \frac{1}{2}$, so

$$C = 1 - \rho - (1 - \rho) \log(1 - \rho) + (1 - \varepsilon - \rho) \log(1 - \varepsilon - \rho) + \varepsilon \log(\varepsilon)$$

5. $I(A, B) = 0.401$ bits, $I(A, C) = 0.125$ bits.

6. (i) Make repeated use of the definition of conditional probability.
    (ii) Use Gibbs' inequality and conditioning to show that $H_B(A) - H_C(A) \le 0$.

7. $H = 1.87$ for *M. lysodeiktus*, $H = 2$ for *E. coli*. As *M. lysodeiktus* has smaller entropy, it is a more complex organism.

9. Use the Kraft – McMillan inequality.

11. Use Theorem 7.4(a).

12. (a) Codeword lengths are 2, 2, 3, 3, 4. A suitable code is 00, 01, 110, 111, 1010.
    (b) Huffman code is 00, 01, 101, 11, 100.

13. The code of (11) has $\eta = 1$, (12(a)) has $\eta = 0.876$, (12(b)) has $\eta = 0.985$.

14. As $S^{(m)}$ consists of m i.i.d random variables, we have

$$H(S^{(m)}) = -\sum_{j_1, \ldots, j_m = 1}^{n} p_{j_1} \cdots p_{j_m} \log(p_{j_1} \cdots p_{j_m})$$

and the result follows via (5.1).

15. (a) Two possible Huffman codes are 0000, 0001, 001, 01, 10, 11 and 0000, 0001, 001, 10, 11, 01.
    As each codeword has the same length, this is effectively two different ways of writing the same optimal code.
    (b) $\xi = 0.019$.

16. $\xi = 0.310$.
19. Use Bayes' rule (Theorem 4.3(a)).
20. $P(E) = 3\varepsilon^2 - 2\varepsilon^3$.
22. Use Lemma 6.1.

Chapter 8

4. (a) $f(x) = \frac{1}{10}(x + 4)$, (b) $\frac{9}{20}$.
5. (i) $c = 1$, (ii) $c = \pi$, (iii) $c = e + 4$.
6. (ii) $F(x) = \frac{x^2}{176}(2x + 3)$, (iii) $\frac{53}{176}$.
8. (i) $\mathbb{E}(X) = \frac{1}{2}(b + a)$, (ii) $\mathrm{Var}(X) = \frac{1}{12}(b - a)^2$.
9. $\mathrm{Var}(X) = \frac{1}{\lambda^2}$.
10. $c = \frac{1}{\pi}$.
11. (i) Substitute $y = \frac{x}{\beta}$ and use $\Gamma(\alpha) = \int_0^\infty y^{\alpha-1}e^{-y}\,dy$, (iv) $\mathbb{E}(X) = \alpha\beta$,
    $\mathrm{Var}(X) = \alpha\beta^2$.
12. (i) Substitute $y = \frac{x^\gamma}{\theta}$, (ii) $\gamma = 1$, (iii) $\mathbb{E}(X^n) = \theta^{n/\gamma}\Gamma(1 + n/\gamma)$.
14. These follows as in the discrete case by using properties of integrals instead of
    sums – particularly linearity

$$\int_a^b (\alpha f(x) + \beta g(x))dx = \int_a^b \alpha f(x)dx + \int_a^b \beta g(x)dx$$

    where α, $\beta \in \mathbb{R}$ and $-\infty \leq a < b \leq \infty$.
15. $\bar{A} = (\mu - c, \ \mu + c)$.
16. Same technique as in the continuous case but using sums instead of integrals.
19. Use Chebyshev's inequality to find

$$P(X > 5) = 0.18(\text{to 2 d.p}'\text{s}).$$

21. (a) 0.09, (b) 0.91, (c) 0.09, (d) 0.91, (e) 0.23, (f) 0.56, (g) 0.
22. You need to estimate $\frac{1}{\sqrt{(2\pi)}}\int_0^1 \exp\left(-\frac{1}{2}z^2\right)\,dz$ by Simpson's rule. To four sig-
    nificant figures this yields 0.3413, which is the same as the result obtained by
    use of tables.
23. Substitute $y = -z$ in the integral for $P(Z \leq 0)$.
25. For $\mathbb{E}(Z)$ use

$$\int z\exp\left(-\frac{1}{2}z^2\right)\,dz = \exp\left(-\frac{1}{2}z^2\right) + C.$$

    For $\mathbb{E}(Z^2)$, integrate by parts writing

$$z^2\exp\left(-\frac{1}{2}z^2\right) = z \cdot z\exp\left(-\frac{1}{2}z^2\right).$$

26. (a) 0.125, (b) 0.75, (c) 0.104.

27. $\mu = 0.181$.

28. Substitute in the integral for the cumulative distribution of Y.

29.

$$G(y) = P(Z \geq y) = \frac{1}{\sqrt{(2\pi)}} \int_y^\infty 1 \cdot \exp\left(-\frac{1}{2}z^2\right) dz$$

$$\leq \frac{1}{\sqrt{(2\pi)}} \int_y^\infty \exp\left(yz - y^2 - \frac{1}{2}z^2\right) dz \text{ by hint.}$$

Now complete the square.

30. (i) $\dfrac{e^{tb} - e^{ta}}{t(b-a)}$.

(ii)

$$M_X(t) = \frac{\lambda}{\lambda - t} \text{ for } -\infty < t < \lambda$$

$$= \infty \text{ if } t \geq \lambda$$

$$\mathbb{E}(X)^n = n!\lambda^{-n}.$$

31. $M_X(t) = (1 - \beta t)^{-\alpha}$.

33. Use a binomial expansion to write

$$\left(1 + \frac{y + \alpha(n)}{n}\right)^n = \sum_{r=0}^n \binom{n}{r} \left(\frac{y + \alpha(n)}{n}\right)^r.$$

Use a similar expansion on $(1 + \frac{y}{n})^n$ and hence deduce that

$$\lim_{n \to \infty} \left\{\left(1 + \frac{y + \alpha(n)}{n}\right)^n - \left(1 + \frac{y}{n}\right)^n\right\} = 0.$$

34. 0.34.

35. $n = 4148$.

36. $p(x, t) = \frac{1}{\sqrt{(2\pi At)}} \exp\left(-\frac{x^2}{2At}\right)$.

37. $H(X) = \log\left(\frac{e}{\lambda}\right)$.

39. (a) Use Exercise 8.38 with $g(x) = \frac{1}{b-a}$ for $x \in (a, b)$.

(b) Use Exercise 8.38 with $g(x) = \lambda e^{-\lambda x}$.

Chapter 9

7. 0.356.

8. (a) $f_W(x, y) = \frac{1}{14}$,   (b) $F_W(x, y) = \frac{1}{14}(x - 3)(y - 2)$,   (c) $\frac{3}{14}$.

9. (a) $\frac{6}{19}$,   (b) 0.

10. (a) $F_W(x, y) = 1 - e^{-x} - e^{-y} + e^{-(x+y)}$.

(b) $f_X(x) = e^{-x}$, $f_Y(y) = e^{-y}$ (so both X and $Y \sim \mathcal{E}(1)$).

(c) Yes.

11. (b) $f_Y(y) = \frac{1}{2}e^{-y/2}$, $f_X(x) = \frac{4}{(4x+1)^2}$.

(c) $f_{Y_x}(y) = \frac{1}{4}y(4x+1)^2 \exp(-\frac{1}{2}y(4x+1))$, $f_{X_y}(x) = 2ye^{-2yx}$.

12. $1 - \frac{1}{2}\exp(-\frac{1}{2}y(4x+1))(2 + y + 4xy)$.

13. (a) $C = 2$

(b) $\frac{1}{4}$ and $\frac{3}{4}$.

(c) $f_X(x) = 2(1-x)$, $f_Y(y) = 2(1-y)$.

(d) $f_{Y_x}(y) = \frac{1}{1-x}$, $f_{X_y}(x) = \frac{1}{1-y}$,

(e) no.

14. $P(X = m) = \frac{\lambda^m e^{-\lambda}}{m!}$, $P(Y = n) = \frac{\mu^n e^{-\mu}}{n!}$. X and Y are Poisson with means λ and μ, respectively, and are clearly independent.

15. 0.0034.

16. The usual business of replacing summation by integration in the earlier proofs.

17. For convenience, we will just do the standard case, so by (9.14),

$$\rho(X, Y) = \text{Cov}(X, Y)$$

$$= \frac{1}{2\pi \sqrt{(1-\rho^2)}} \int_{-\infty}^{\infty} \int_{-\infty}^{\infty} xy$$

$$\times \exp\left\{-\frac{1}{2(1-\rho^2)}(x^2 - 2\rho xy + y^2)\right\} dxdy$$

$$= \frac{1}{2\pi \sqrt{(1-\rho^2)}} \int_{-\infty}^{\infty} \int_{-\infty}^{\infty} x \exp\{-\frac{1}{2(1-\rho^2)}(x - \rho y)^2\}$$

$$.y \exp\left\{-\frac{1}{2}y^2\right\} dxdy.$$

Now substitute $z = \frac{x-\rho y}{\sqrt{(1-\rho^2)}}$ to find that

$$\rho(X, Y) = \mathbb{E}(Y)\mathbb{E}(\sqrt{(1 - \rho^2)}X) + \rho\mathbb{E}(Y^2) = \rho.$$

By Example 9.10, we know that X and Y are independent if and only if $\rho = 0$, that is, $\text{Cov}(X, Y) = 0$ but $\text{Cov}(X, Y) = \mathbb{E}(XY) - \mathbb{E}(X)\mathbb{E}(Y)$ and the result follows.

18. Use Lemma 9.2 in (9.18).

19. (a) $\mathbb{E}(\chi_A(X)) = \int_a^b \chi_A(x)f(x)dx = \int_A f(x)dx = p_X(A)$.

21. Use the convolution formula of the preceding exercise, noting that you only need to integrate from 0 to z (Why?)

22. $X + Y \sim N((\mu_1 + \mu_2), (\sigma_1^2 + \sigma_2^2))$.

23. $\mathbb{E}_x(Y) = \mathbb{E}(Y_x) = \mu_2 + \rho(x - \mu_1)$ by Example 9.12.

26. Replace all the sums in the proof of Theorem 6.5 by integrals.

27. $H(X, Y) = \log(2\pi e \sigma_1 \sigma_2(1 - \rho^2)^{1/2})$.

28. (a) The identity follows from (9.23) and Exercise 9.27 above.
   (b) $I(X, Y) \geq 0$ follows by using Gibbs' inequality (Exercise 9.38) to imitate the argument of Exercise 6.10.
30. (a) $H(X) = \log(V_2 - V_1)$, $H(Y) = \log(W_2 - W_1)$.
   (b) $H(X, Y) = \log((V_2 - V_1)(W_2 - W_1))$.
   (c) $I(X, Y) = 0$ (input and output are independent so no communication is possible).
31. (a) $H(Y) = -0.28$ bits, (b) $H_X(Y) = -0.72$ bits, (c) $I(X, Y) = 0.44$ bits.

Chapter 10

1. (a) 0 (b) 0.036 (c) 0.192.
2. Both equal 0.12.
3. Assume that the random variables are all discrete, then by independence

$$P(S_n = x_n | S_1 = x_1, S_2 = x_2, \ldots, S_{n-1} = x_{n-1})$$

$$= P(X_n = x_n - x_{n-1})$$

$$= P(S_n = x_n | S_{n-1} = x_{n-1}).$$

5. (a) $\frac{1}{2}$ (b) $\frac{1}{4}$ (c) 0 (d) $\frac{1}{8}$ (e) $\frac{1}{16}$.
6. (a) $(1 - \alpha)^2$ (b) $(1 - \alpha)^2 + \alpha^2$
7. (b) 0.13, (c) 0.16.
8. $\pi_0^{(2)} = 0.42$, $\pi_0^{(3)} = 0.416$.
9. (a) 0.006, (b) (i) 0.29 (ii) 0.231, (iii) $\pi^{(2)} = (0.481, 0.237, 0.282)$.
10. $\frac{1}{2}[1 + (1 - 2\alpha)^5]$

11. (a) $P = \begin{pmatrix} 0 & p & 0 & 0 & 1-p \\ 1-p & 0 & p & 0 & 0 \\ 0 & 1-p & 0 & p & 0 \\ 0 & 0 & 1-p & 0 & p \\ p & 0 & 0 & 1-p & 0 \end{pmatrix}$, (b) 0.001 62.

12. 0.466.
14. (a) $(\frac{4}{7}, \frac{3}{7})$, (b) $(\frac{1}{3}, \frac{1}{3}, \frac{1}{3})$, (c) $(\frac{1}{25}, \frac{18}{25}, \frac{6}{25})$.
23. (a) 0.9197, (b) 1.489, (c) 0.8172.

Index